Small for Gestational Age. Causes and Consequences

Pediatric and Adolescent Medicine

Vol. 13

Series Editors

Wieland Kiess Leipzig

David Branski Jerusalem

Small for Gestational Age

Causes and Consequences

Volume Editors

Wieland Kiess Leipzig
Steven D. Chernausek Oklahoma City, Okla.
Anita C.S. Hokken-Koelega Rotterdam

23 figures, 7 in color, and 17 tables, 2009

KARGER   Basel · Freiburg · Paris · London · New York · Bangalore ·
Bangkok · Shanghai · Singapore · Tokyo · Sydney

Prof. Wieland Kiess, MD
Hospital for Children and Adolescents
University of Leipzig
Leipzig, Germany

Prof. Steven D. Chernausek, MD
Children's Medical Research Institute (CMRI)
Diabetes and Metabolic Research Program
Department of Pediatrics
University of Oklahoma Health Sciences Center, Oklahoma City, Okla., USA

Prof. Anita C.S. Hokken-Koelega, MD, PhD
Division of Endocrinology
Department of Paediatrics, Sophia Children's Hospital
Erasmus University Medical Center, Rotterdam
The Netherlands

Library of Congress Cataloging-in-Publication Data

Small for gestational age : causes and consequences / volume editors, Wieland
Kiess, Steven D. Chernausek, Anita C.S. Hokken-Koelega.
p. ; cm. -- (Pediatric and adolescent medicine, ISSN 1017-5989 ; v. 13)
Includes bibliographical references and index.
ISBN 978-3-8055-8657-3 (hard cover : alk. paper)
1. Fetal growth retardation. 2. Birth weight, Low. I. Kiess, W. (Wieland)
II. Chernausek, Steven D. III. Hokken-Koelega, Anita C.S. IV. Series.
[DNLM: 1. Fetal Growth Retardation--physiopathology. 2. Infant, Small for
Gestational Age. 3. Fetal Growth Retardation--etiology. W1 PE163HL v.13 2008 /
WS 420 S635 2008]
RG629.G76.S63 2009
618.3'2--dc22
2008031195

© Copyright 2009 by S. Karger AG, P.O. Box, CH–4009 Basel (Switzerland)
www.karger.com
Printed in Switzerland on acid-free and non-aging paper (ISO 9706) by Reinhardt Druck, Basel
ISSN 1017–5989
ISBN 978–3–8055–8657–3

Contents

Preface

For medical, ethical, socioeconomic and humanitarian reasons, it is mandatory to foster research into causes and consequences of intrauterine growth retardation in the human. While a lot is known about the causes and consequences of being born small for gestational age, even more has yet to be discovered until the ethical, socioeconomic and health challenges and dilemmas of low birth weight have been dealt with. This book aims to address the most urgent questions and most relevant issues in terms of the clinical care of small for gestational age infants. In addition, frontiers of research into the causes and consequences of being born small for gestational age are being discussed.

H. Wollmann from the University of Tübingen, Germany, has worked in the area of pediatric endocrinology and neonatology both as a clinician and as a clinician scientist for more than two decades. He sets the stage with his chapter on definitions and the etiology of being born small for gestational age. In their chapter on the diagnosis and management of in utero growth failure, Odibo and co-workers from St. Louis, Mo., USA, delineate the prenatal etiology and manifestations of growth retardation. They point out that therapeutic measures during intrauterine life would be and will be the ultimate prevention and treatment of all disorders leading to small-for-gestational age birth. The origins of adult metabolic disorders are thought to be found in fetal growth restriction. In their chapter, Krechowec and co-workers report on their fascinating studies with regard to the developmental origins of adult diseases after fetal growth restriction. Chernausek from the University of Oklahoma, USA, draws attention to the wealth of knowledge that today abounds in relation to molecular genetic disorders of fetal growth. He is among the most successful researchers in this area of science and an experienced pediatric endocrinologist involved in the care of SGA children. Harder's group from Berlin, Germany, discuss the hypothesis of the developmental/fetal origins of adult disease in a most elegant way. Franke and co-workers from New Zealand draw our attention to the global perspective of low birth weight: it is very clear today that optimal fetal development ensures good postnatal health. In a short chapter, the group from Leipzig, Germany, summarizes our knowledge on the role of the GH/IGF system in respect to fetal growth. Knüpfer, Leipzig, Germany, suggests practical algorhythms for the clinical management of small-for-gestational-age babies from the neonatologist's point of view. Dahlgren, Göteborg, Sweden, writes on the management of short stature in SGA children, while Hokken-Koelega from the Sophia Children's Hospital, Rotterdam, The Netherlands, reports on her very extensive experience in the regulation of puberty and adrenarche in SGA children. Tuvemo and co-workers from Uppsala, Sweden, summarize our knowledge of neurological and intellectual consequences of being born small for gestational age. This aspect has very long range sequelae and is of utmost importance. It is not just growth, metabolic and cardiovascular consequences that make life for SGA children special but also their development that raises concern and needs attention. Finally, Qadir's group from Karachi, Pakistan, give deep insight into the causes and consequences of low birth weight in developing countries. Interventions that affect maternal and child undernutrition and nutrition-related outcomes include promotion of breastfeeding; strategies to promote complementary feeding, with or without provision of food supplements; micronutrient interventions; general supportive strategies to improve family and community nutrition, and reduction of disease burden.

The editors wish to thank Dr. Thomas Karger and his devoted staff at S. Karger, Publishers, Basel, Switzerland, for making this volume possible. We all wish to dedicate this book to Steven Karger who sadly was not able to see it in its finished state. We hope that the readers will gain new insights into the causes and consequences of small-for-gestational-age children and draw strength for the care of babies from their new knowledge and insights.

Wieland Kiess, Leipzig
Steven D. Chernausek, Oklahoma City, Okla.
Anita C.S. Hokken-Koelega, Rotterdam

Kiess W, Chernausek SD, Hokken-Koelega ACS (eds): Small for Gestational Age. Causes and Consequences.
Pediatr Adolesc Med. Basel, Karger, 2009, vol 13, pp 1–10

Children Born Small for Gestational Age: Definitions and Etiology

Hartmut A. Wollmann

Paediatric Endocrinology Section, University Children's Hospital, Eberhard Karls University,
Tübingen, Germany; Pfizer Endocrine Care, Tadworth, Surrey, UK

Abstract

Children born small for gestational age by pediatricians are usually defined by birth weight and/or length below -2 SD. Perinatally, definitions vary and methods including physiological factors influencing birth measures are advantageous with maternal height, maternal weight in early pregnancy, parity and ethnic origin as well as the sex of the baby as the most important ones. The etiology of SGA covers a broad spectrum of maternal, environmental, placental and fetal factors, but still in a significant proportion of cases the reason for reduced birth measures remains unclear. Quantitatively, maternal hypertension and smoking are important, whereas other etiologies frequently cause a more severe damage to the fetus. In the future, the specific effect of different etiologies on long-term outcome needs to be worked out.

Copyright © 2009 S. Karger AG, Basel

Definition of Small for Gestational Age

Historically, in the 1950s of the last century, when the term 'intrauterine growth retardation' (IUGR) was beginning to be used, the knowledge about these children concerning causes, prenatal and postnatal behavior and their final outcome was limited. At that time, they were usually labeled 'premature' [1]. Meanwhile, the term 'small for gestational age' (SGA) is widely used [2] with a considerable ongoing debate on how to define that properly, depending on whether the intrauterine, the perinatal or the childhood situation is considered. As IUGR/SGA is associated with significant antenatal as well as postnatal morbidity, which goes far beyond short stature [3], an accurate definition is essential for optimal care.

Prerequisites for an accurate definition of SGA are the following [2]: (1) gestational age should be known, best based on the date of conception or on a first trimester ultrasound examination, (2) measurements of birth weight, birth length and head circumference should be performed by trained personal, following standardized, appropriate procedures, and (3) the results should be compared with population-specific, updated reference tables, allowing to specifically classify children as appropriate for gestational age (AGA) or SGA, according to the cut-off chosen.

During perinatal care, the 10th percentile for body weight is used most frequently to define a newborn as SGA, as reliable measurement of body length immediately after birth is rare and is not even performed in some countries. Using this relatively broad definition, most of those children presenting with perinatal problems are recorded. However, it should be emphasized that even a baby born with birth measurements at the 10th percentile may have suffered from severe intrauterine growth retardation/restriction (IUGR) late in pregnancy, eventually crossing from the 90th to the 10th percentile during the 3rd trimester due to severe placental malfunction.

For growth and long-term outcome, the SGA consensus conference [2] recommended to define a newborn as SGA if birth weight and/or birth length is less than $-2\,SD$ for gestational age, using population-specific standards. Basically, this definition is not linked to pathology, it includes the whole spectrum from short normals (constitutional smallness) as well as IUGR infants. This definition has meanwhile been accepted by regulatory authorities in the USA and Europe. Based on this definition (birth length and/or birth weight), approximately 5% of all newborns will be labeled SGA and deserve specific attention and follow-up. Up to 90% of them will exhibit early and rapid catch-up growth with normalization of length and weight.

SGA babies can be subdivided [4], depending on whether primarily birth weight (SGA_{weight}), birth length (SGA_{length}), or both birth measures are reduced. This subclassification has implications for the long-term outcome as, for example, postnatal catch-up growth is relatively frequent in SGA_{weight} whereas the babies with reduced weight and length have a high risk to end up short.

Intrauterine Growth Retardation
The term IUGR, even if it is frequently used synonymously with SGA, describes a subgroup of small babies with proven intrauterine growth restriction. This term should only be used if either, a reduced fetal growth velocity can be documented by repeated prenatal ultrasound measurements and/or a specific cause for can be identified (e.g. restricted placental fuel supply, defined syndromes, toxic effects or fetal infections). Thus, a child who is born SGA frequently has not suffered from IUGR, and infants born after a short period of IUGR are not necessarily SGA [5].

Definition in Premature Infants
In very low birth weight premature babies (VLBW), differentiation between SGA and AGA status may be difficult due to the lack of reliable references for healthy premature infants, problems with exact definition of gestational age and their specific

growth pattern [6]. Even if they are born AGA, a subgroup of them exhibit postnatal but preterm growth restriction with a growth pattern very similar to VLBW SGA babies, which even exceeds perinatal life and was documented until the age of 5 years. Therefore, with respect to long-term growth, differentiation of SGA and AGA in VLBW premature babies is possibly of minor importance.

Customized Growth Charts

During recent years, 'customized' fetal growth curves and birth weight percentiles have been proposed, adjusting growth standards to physiological factors known to affect fetal growth, i.e. birth weight and length. Among these variables, maternal height, maternal weight in early pregnancy, parity and ethnic origin as well as the sex of the baby turned out to be the most important ones [7]. Not surprisingly, paternal height has an impact as well, but is of minor importance compared to maternal height.

This method allows one to reduce false-positive results of fetal growth assessments and helps to identify those babies who are pathologically small. Even modest growth restriction (i.e. between the 3rd and 10th percentile) can be identified as pathological with this method. For neonatal endpoints, like neonatal morbidity and perinatal death, this method is able to predict the risk for an individual child more precisely, compared to population-based standards [8]. However, for long-term heath outcomes, like short stature in SGA and metabolic disturbances later in life, the benefit of using customized growth charts still needs to be demonstrated.

Definition during Childhood

Those SGA children (approximately 10%) who do not exhibit catch-up growth during the first 2 years of life, will remain short throughout childhood and have an increased risk for short stature as adults [4]. As meanwhile treatment with recombinant human growth hormone (rhGH) is a viable therapeutic option for these children, in addition to the neonatal SGA definition a childhood definition is needed to qualify children for treatment.

For the USA, the height cut-off at the start of treatment is not specified, therefore the general cut-off for definition of short stature (height < -2 SD) is normally used for SGA children. In the EU, a specific stricter height cut-off of < -2.5 SD (0.6th percentile) has been approved for the treatment with GH, restricting the potential number of patients that could be treated. In addition, catch-up growth has to be excluded by demonstrating a height velocity below the mean at start of treatment. Furthermore, height of the patient at start of treatment has to be at least 1 SD below target height, to exclude patients with familial short stature from treatment.

Table 1. Physiological factors influencing birth weight

Maternal factors
Maternal height
Pre-pregnancy weight
Low weight gain
Living at high altitude
Parity
Delivery at age <16 or >35
Ethnicity
Previous history of SGA
Paternal factors
Paternal height
Fetal factors
Sex

Etiology of Small for Gestational Age

Physiological Factors Influencing Birth Weight

Per definition, SGA covers short normal children, as well as those with proven IUGR, i.e. proven pathology. Therefore, discussing etiology requires a discussion of physiological factors influencing birth measures in general, as well as pathological factors directly or indirectly restricting fetal growth.

Table 1 summarizes the most important physiological factors influencing birth weight and length.

The important influence of maternal anthropometry before and during pregnancy is illustrated by a prospective investigation of 1,104 pregnant women, followed from the 3rd to the 5th month of pregnancy until delivery [13]. In general, the best predictors of birth weight turned out to be weight at registration (i.e. between months 3 and 5 of pregnancy) and weight at month 9, reflecting early pregnancy (or even prepregnancy) weight and weight gain during pregnancy. Each 1 kg increase in body weight at registration was associated with an increase of birth weight of about 260 g. Height at registration – after controlling for weight – turned out to be significantly related to birth weight as well; however, the impact of height on birth weight of the baby was less important than that of body weight.

In contrast, maternal overweight or obesity before pregnancy is associated with an increased risk of foetal macrosomia and perinatal mortality [15, 16], but it protects the child from SGA birth, the children of overweight or obese mothers tend to be rather heavy or even macrosomic.

There is limited information about paternal influences on birth weight. In a prospective study, Wilcox et al. [12] investigated this influence of paternal anthropometric measures controlling for gestational age, ethnicity, maternal height and other confounding factors. When considered in isolation, both paternal height and weight were significantly positively associated with birth weight. As, however, paternal size is correlated with maternal size and smoking habits, only paternal height had a significant impact on birth weight: If the partner of an average woman is short ($<-2\,SD$), then on average the baby will be 183 g lighter than if he is tall. This is probably a genetic effect, with paternal genes expressed in the placenta undergoing restriction by maternal control protecting the mother from the risk associated with a too heavy/large baby. Overall, the effects of paternal size are considerably smaller than those of maternal size, expressing the much stronger effect of mothers' size and nutritional support to the intrauterine environment.

There is ample evidence [14] that racial or ethnic origin has an impact on birth weight. In most cases, white, black and Hispanic groups in the USA were compared and in most cases, the white population tended to have higher birth weights. However, most of the studies lack the methodological rigor to exclude socioeconomical confounders. Therefore, the question whether ethnic origin – in addition to mother's height – significantly influences birth weight is still unanswered. This is underscored by a large USA study, comparing birth weights of large cohorts of USA-born white women, USA-born black women and African-born black women. These data indicate that social rather than genetic factors appear to be the chief contributors to low birth weight and the associated increase in infant mortality [17].

Infants born to primiparous mothers have lower mean birth weights and higher rates of SGA birth compared to multiparous, tall or heavy mothers [15]. However, in contrast to the influence of maternal short stature or low pre pregnancy body mass index, the reduced growth of fetuses born to primiparous mothers is associated with a higher risk of perinatal death, indicating that the impact of parity on birth weight is a pathological one. This finding puts questions to the use of customized growth standards, as variation of these 'physiological' factors should not affect risks of mortality and short- and long-term morbidity [15].

Young maternal age confers a considerable risk of adverse outcomes of pregnancy, including low birth weight and prematurity [16]. This may be caused by a functional immaturity of the reproductive system of the teenage mother, as well as deleterious social environment and genetic influences. Fraser et al. [16] studied 134,000 girls and women aged 13–24 years who delivered singleton, first-born children. The relative risk to have a low-birth-weight infant was found to be increased 2-fold if the mother was younger than 17 years. In addition, the risk for premature delivery was elevated as well. This elevation was independent of relevant sociodemographic factors like marital status, level of education and adequacy of prenatal care. Advanced maternal age is an independent risk factor for IUGR [23]. At maternal age of 40 years or older, the odds ratio for IUGR in this study was 3.2 (CI 1.9–5.4) based on that a screening for IUGR in pregnant women age 35 years or older was recommended.

The interpregnancy interval among multiparous women is directly associated with birth weight. The greatest risk for SGA birth was observed in women with the shortest interpregnancy intervals [18]. The specific mechanisms responsible for this association are unknown. Quantitatively, however, the effect of the interpregnancy interval on fetal growth is relatively small compared to that of a prior history of a low-birth-weight infant. Among the maternal factors, a prior SGA child leads to a reduction of approximately 10% (340 g) in birth weight, an effect comparable to that of pregnancy-associated hypertension (−10%) or maternal diabetes (+10%) [19]. The mechanism again is unclear, but most probably genes impacting on growth and function of the uterus and the placenta as well as environmental factors might be causative.

Perinatal outcomes, such as preterm delivery, low birth weight and some obstetric complications, are increased significantly after in vitro fertilization (IVF) compared with spontaneously conceived pregnancies. In a retrospective study, investigating 2,956 pregnancies, 30.8% were found to be multiple and 40.6% of the neonates had a birth weight below 2,500 g. A birth weight under the 10th percentile of the national reference was found in 16.7%. Analyzing the singleton pregnancies separately revealed lower numbers [20]. The authors conclude that the increased risk for SGA birth as well as the preterm delivery in 29.2% was primarily caused by the high proportion of multiple pregnancies. Latest results [21] confirm this observation by comparing neonatal outcomes of single-embryo transfer in IVF/ICSI with spontaneously conceived babies providing almost identical values for birth weight, gestational age and the rate of prematurity.

Maternal Factors Associated with Impaired Fetal Growth

Maternal diseases, the impact of socioeconomic status, and drug abuse are summarized in table 2.

Maternal causes restricting fetal growth are multiple [5, 10, 21], including a long list of medical conditions, infections like toxoplasmosis, rubella, cytomegalovirus, herpesvirus, malnutrition, smoking and the abuse of illicit drugs, and most of them are discussed in detail elsewhere.

Quantitatively, pregnancy-associated hypertension, eventually complicated with preeclampsia, is the most important maternal factor influencing fetal growth: severe, pregnancy-induced hypertension reduces average birth weight by approximately 10%. Interestingly, a pre-existing uncomplicated maternal hypertension does not have a comparable effect on birth weight.

Among the environmental factors, cigarette smoking by the mother during pregnancy is by far the most important cause of restricted intrauterine growth. Based on a meta-analysis, Kramer [14] concluded that in developed countries fetal growth retardation in up to 40% is caused by cigarette smoking of the mother. There is a direct, linear correlation between the number of cigarettes smoked per day and the degree of growth retardation [24]. The percentage of low-birth-weight infants is increase 3-fold if the mother smokes more than 20 cigarettes per day. Smoking one pack of cigarettes

Table 2. Maternal factors associated with impaired fetal growth: adapted from Bryan and Hindmarsh [9] and Saenger et al. [3]

Maternal medical complications
Preeclampsia
Acute or chronic hypertension
Antepartum hemorrhage
Severe chronic disease
Severe chronic infections
Systemic lupus erythematosus
Antiphospholipid syndrome
Anemia
Malignancy
Abnormalities of the uterus
Uterine fibroids

Maternal social conditions
Malnutrition
Low socioeconomic status
Drug use
　Smoking
　Alcohol
　Other drugs

per day reduces average birth weight by 5%; in addition, the abortion rate in early pregnancy and the risk of prematurity in late gestation are significantly elevated. Other influencing factors discussed above like male sex of the fetus, prepregnancy weight, pregnancy weight gain and maternal height are of minor importance compared to smoking. Passive smoking [25] does not result in a significant effect on birth weight, as is true for snuffed nicotine, indicating that other components of cigarette smoke besides nicotine are causative for reduced fetal growth. Still, preventing women planning pregnancy and pregnant women from smoking is by far the most effective method of reducing the number of SGA infants in developed countries. For the developing countries, maternal nutrition is of comparable importance.

Compared to smoking, the effect of other drugs (alcohol, illicit drugs) is quantitatively of minor importance for the affected fetus [26]. However, due to the association with congenital malformations and the adverse long-term outcome of the children, the abuse of any of these drugs is extremely deleterious. This was highlighted in a recent, prospective, population-based survey on fetal alcohol syndrome: 64% of the children had low birth weight, 36% were preterm and 78% were exposed to one or more additional drugs [27].

Of these children, 53% had microcephaly and 86% had evidence of central nervous dysfunction. As there is no safe threshold level of alcohol consumption during pregnancy and probably most of the children with minor effects (fetal alcohol effects) are not diagnosed, complete abstinence during pregnancy must be recommended.

Table 3. Fetal and placental factors associated with impaired fetal growth: adapted from Bryan and Hindmarsh [9]

Fetal problems
Multiple births
Malformation
Chromosomal abnormalities
Inborn errors of metabolism
Intrauterine infections

Abnormalities of the placenta
Reduced blood flow
Reduced area for exchange
Infarcts
Hematomas
Partial abruption

Coffee consumption is widespread in western populations, as well as in some developing countries, and it is consumed by pregnant women as well. Caffeine has well-described pharmacological effects and easily crosses the placenta. Due to the fetus' relative inability to metabolize the substance, pregnancy outcome may be effected by caffeine consumption of the mother. In spite of that, published data are not homogeneous. Whereas older publications state that 'at worst caffeine may have relatively subtle adverse consequences for human reproduction' [28], more recent data show that results suggest that a high caffeine intake in the third trimester may be a risk factor for fetal growth retardation, in particular if the fetus is a boy. The risk of SGA birth was nearly doubled if the mother had a high rather than a low caffeine intake in the third trimester. The reason for the gender difference in caffeine sensitivity is currently unknown [29].

Fetal and Placental Factors Associated with Impaired Fetal Growth

A short summary of fetal and placental factors associated with impaired fetal growth is listed in table 3. Compared to other etiologies, fetal conditions causing IUGR are rare, but are frequently associated with severe growth restriction as well as other malformations and dysmorphic features and a poor prognosis with respect to long-term outcome.

Overall, fetal infections contribute to the growth restriction of the fetus in 5–10% of the cases. A direct causal relationship between congenital infections disease and IUGR is proven for two viruses, rubella and cytomegalovirus (CMV) [30]. A possible relationship exists for the varicella-zoster virus (VZV) and human immunodeficiency virus. As the association of maternal/fetal viral infection and IUGR is frequently associated with a poor prognosis and effective therapeutic interventions do not exist, prevention of infection is the most important approach. In case of rubella, immunization programs have been quite successful, whereas in other cases – where no immunization exists – termination of pregnancy in case of proven infection and an affected fetus has been recommended.

Placental factors influencing fetal growth are ample: from structural, histological and cytogenetic abnormalities, maternal and fetal perfusion, disturbed substrate transfer, effects of disturbed angiogenesis, growth factors, immune-mediated disturbances, etc. [31]. The correlation between disturbed placental function and fetal growth restriction is very strong, irrespective of the specific cause. However, the mechanisms underlying the effect of placental dysfunction on fetal growth are still poorly understood.

References

1 Warkany J, Monroe BB, Sutherland BS: Intrauterine growth retardation. Am J Dis Children 1961;102:127–157.

2 Clayton PE, Cianfarani S, Czernichow P, Johannsson G, Rapaport R, Rogol A: Management of the child born small for gestational age through to adulthood: a consensus statement of the International Societies of Pediatric Endocrinology and the Growth Hormone Research Society. J Clin Endocrinol Metab 2007;92:804–810.

3 Saenger P, Czernichow P, Hughes I, Reiter EO: Small for gestational age: short stature and beyond. Endocr Rev 2007;28:219–251.

4 Karlberg J, Albertsson-Wikland K: Growth in full term small-for-gestational-age infants: from birth to final height. Pediatr Res 1995;38:733–739.

5 Lee PA, Chernausek SD, Hokken-Koelega AC, Czernichow P: International Small for Gestational Age Advisory Board consensus development conference statement: management of the short child born small for gestational age. Pediatrics 2001;111:1253–1261.

6 Finken MJ, Dekker FW, de Zegher F, Wit JM: Long-term height gain of prematurely born children with neonatal growth restraint: parallelism with the growth pattern of short children born small for gestational age. Pediatrics 2006;118:640–643.

7 Gardosi J: Customized fetal growth standards: rationale and clinical application. Semin Perinatol 2004; 28:33–40.

8 Gardosi J: New Definition of small for gestational age based on fetal growth potential. Horm Res 2006; 65(suppl 3):15–18.

9 Bryan SM, Hindmarsh PC: Normal and abnormal fetal growth. Horm Res 2006;65(suppl 3):19–27.

10 Wollmann HA: Intrauterine growth restriction: definition and etiology. Horm Res 1998;49(suppl 2):1–6.

11 Hindmarsh PC, Geary MP, Rodeck CH, Kingdom JC, Cole TJ: Interauterine growth and its relationship to size and shape at birth. Pediatr Res 2002;52:263–268.

12 Wilcox MA, Newton CS, Johnson IR: Paternal influences on birthweight. Acta Obstet Gynecol Scand 1995;74:15–18.

13 Nahar S, Mascie-Taylor CG, Begum HA: Maternal anthropometry as a predictor of birth weight. Publ Health Nutr 2007;10:965–970.

14 Kramer MS: Determinants of low birth weight: methodological assessment and meta-analysis. Bull World Health Org 1987;65:663–737.

15 Zhang X, Cnattingius S, Platt RW, Joseph KS, Kramer MS: Are babies born to short, primiparous, or thin mothers 'normally' or 'abnormally' small? J Pediatr 2007;150:603–607.

16 Fraser AM, Brockert JE, Ward RH: Association of young maternal age with adverse reproductive outcomes. N Engl J Med 1995;332:1113–1117.

17 David RJ, Collings JW: Differing birth weight among infants for US-born Blacks, African-born Blacks, and US-born Whites. N Engl J Med 1997; 337:1209–1214.

18 Lieberman E, Lang JM, Ryan KJ, Monson RR, Schoenbaum SC: The association of inter-pregnancy interval with small for gestational age births. Obstet Gynecol 1989;74:1–5.

19 Kramer MS, Olivier M, McLean FH, Dougherty GE, Willis DM, Usher RH: Determinants of fetal growth and body proportionality. Pediatrics 1990;86:18–26.

20 Koudstaal J, van Dop PA, Hogerzeil HV, Kremer JA, Naaktgeboren N, van Os HC, Tiemessen Ch, Visser GH: Pregnancy course and outcome in 2,956 pregnancies after in-vitro fertilization in Netherlands. Ned Tijdsch Geneeskd 1999;143:2375–2380.

21 De Neubourg D, Gerris J, Mangelschots K, Van Royen E, Vercruyssen M, Steylemans A, Elseviers M: The obstetrical and neonatal outcome of babies born after single-embryo transfer in IVF/ICSI compares favourably to spontaneously conceived babies. Hum Reprod 2006;21:1041–1046.

22 Pollack RN, Divon MY: Intrauterine growth retardation: definition, classification, and etiology. Clin Obstet Gynecol 1988;35:99–107.

23 Odibo AO, Nelson D, Stamilio DM, Sehdev HM, Macones GA: Advanced maternal age is an independent risk factor for intrauterine growth restriction. Am J Perinatol 2006;23:325–328.

24 Stillman RJ, Rosenberg MJ, Sachs BP: Smoking and reproduction. Fertil Steril 1986;46;545–566.

25 Steyn K, de Wet T, Saloojee Y, Nel H, Yach D: The influence of maternal cigarette smoking, snuff use and passive smoking on pregnancy outcomes: the Birth To Ten Study. Paediatr Perinat Epidemiol 2006;20:90–99.

26 Friedman JM: Effects of drugs and other chemicals on fetal growth. Growth Genet Horm 1992;8:1–5.

27 Elliott EJ, Payne JM, Morris A, Haan E, Bower CA: Fetal alcohol syndrome: a prospective national surveillance study. Arch Dis Child 2007;17.

28 James JE, Paul I: Caffeine and human reproduction. Rev Environ Health 1985;5:151–167.

29 Vik T, Bakketeig LS, Trygg KU, Lund-Larsen K, Jacobsen G: High caffeine consumption in the third trimester of pregnancy: gender specific effects on fetal growth. Paediatr Perinat Epidemiol 2003;17:324–331.

30 Kilby M, Hodgett S: Perinatal Viral Infections as a Cause of Intrauterine Growth Restriction; in Kingdom J, Baker P (eds): Intrauterine Growth Restriction; Aetiology and Management. London, Springer, 1999, pp 29–47.

31 Fox H: Placental Pathology; in Kingdom J, Baker P (eds): Intrauterine Growth Restriction; Aetiology and Management. London, Springer, 1999, pp 187–201.

Prof. Hartmut A. Wollmann, MD, PhD
Pfizer Endocrine Care
Wilhelmstrasse 44
DE–72074 Tübingen (Germany)
Tel. +49 7071 256 883, Fax +49 7071 256 884, E-Mail hartmut.wollmann@pfizer.com

Kiess W, Chernausek SD, Hokken-Koelega ACS (eds): Small for Gestational Age. Causes and Consequences.
Pediatr Adolesc Med. Basel, Karger, 2009, vol 13, pp 11–25

Diagnosis and Management of in utero Growth Failure

Anthony O. Odibo[a] · Yoel Sadovsky[a,b]

Departments of [a]Obstetrics and Gynecology and [b]Cell Biology and Physiology,
Washington University School of Medicine, St. Louis, Mo., USA

Abstract

Although fetal growth restriction (FGR) is a relatively common obstetric disease, our understanding of
the etiology and pathophysiology of this condition is relatively rudimentary. Moreover, once FGR is diagnosed, there are no known interventions that cure or definitively ameliorate the disease. This chapter
focuses on the diagnostic and management paradigms in pregnancies complicated by FGR, emphasiz-
ing challenges that result from overlapping classifications, and delineating biophysical parameters currently used to diagnose FGR and characterize its severity. Several molecular tools that might aid in the
definition of FGR are reviewed. Finally, we suggest pathways that may be useful in the clinical management of pregnancies complicated by FGR.

Copyright © 2009 S. Karger AG, Basel

The term small for gestational age (SGA) implies that the fetus is smaller than its
expected size. It has been historically defined as a neonatal weight of less than 2,500 g
at term [1]. More modern definitions use an estimated weight below the 10th percentile for the gestational age or weight that is less than 2 SD below the anticipated
value for the gestational age [2]. Because of its clinically relevant consequences, some
authors prefer to define SGA as weight below the 5th or even 3rd percentile [3]. As it
is impossible to determine the expected growth rate for each fetus, gestational age-
adjusted reference standards are used to define SGA. A fraction of fetuses with SGA
exhibit a pathological growth restriction. In contrast to SGA, fetal growth restriction
(FGR) implies a failure of the fetus to achieve its anticipated growth potential. This
likely reflects an underlying abnormality that impedes the fetus from normal growth.
The distinction between the general classification of SGA and the more specific FGR
is important, because FGR is the second leading contributor to perinatal mortality.
Irrespective of its etiology, the syndrome of FGR is generally associated with greater
than three-fold increase in perinatal mortality and morbidity, largely attributed to a
higher incidence of intrauterine fetal demise, intrapartum fetal morbidity and a
greater need for operative deliveries, including cesarean sections. A higher rate of

immediate morbidity and mortality characterizes newborns with FGR, related primarily to neonatal acidosis and its sequelae: hypoglycemia, polycythemia, hyperbilirubinemia and hypothermia [4, 5]. While the insult occurs in utero, the deleterious influence of FGR may linger into childhood and is associated with significant long-term developmental delay and neurobehavioral dysfunction. Importantly, work by Barker et al. [6] and others have suggested a link between substandard embryonic growth and metabolic disorders during adult life, including insulin resistance, atherosclerosis and their sequelae.

Diagnosis of FGR

Identification of the population at-risk for FGR is paramount to early diagnosis of this disease. Moreover, determining the etiology of FGR may assist in optimizing the nature and timing of clinical interventions designed to attenuate the adverse impact on the fetus. Factors that influence fetal growth can be intrinsic to the fetus or represent extrinsic insults that stem from the maternal environment. The most common intrinsic causes of FGR are fetal anomalies (aneuploidy or other), multifetal gestation and congenital infections (e.g. rubella, CMV, toxoplasmosis) [7]. Extrinsic factors are more common, and are frequently suspected based on the maternal history. Common examples include preeclampsia and other hypertensive disorders, maternal vascular diseases, diabetes and thrombophilia, which tend to cause utero-placental hypoperfusion, likely reflecting inadequate invasion of extravillous trophoblast into the decidua and inner myometrium. Other extrinsic causes include hypoxia from severe cardiac disease, pulmonary disease or living in high altitude, profound anemia (particularly hemoglobinopathy), severe malnutrition, or exposure to tobacco, drugs, toxins, teratogens or medications that can affect uterine blood flow or fetal growth. Many of these disorders affect fetal growth through their impact on the placenta, causing chronic abruption, infarcts, chronic villitis, previllous fibrin or fetal thrombotic vasculopathy [8–14]. Lastly, it is not uncommon to diagnose FGR in the absence of any known intrinsic or extrinsic risk factor. Some of the cases of idiopathic FGR exhibit the characteristic histo-pathological changes that are associated with extrinsic conditions. In more than 20% of fetuses with idiopathic FGR, the diagnosis can be attributed to confined placental mosaicism, where at least two cell lines with different chromosomal complements are present in the placenta [15]. In the absence of risk factors, a history of a previous fetus with FGR may be the only hint to the possibility of recurrent FGR [7, 16].

Clinical Diagnosis
The determination of the most accurate gestational age by the last menstrual period and sonography (see below) is crucial for evaluation of fetal growth. The first indication of fetal growth abnormality is frequently subnormal uterine size, determined by

clinical impression and supported by abdominal palpation and a direct measurement of the symphyseal-fundal distance [17]. A suspicion of abnormal fetal growth is increased by the presence of the risk factors discussed earlier, or in the setting of poor maternal weight gain.

Role of Sonography

Whereas routine sonographic biometry may frequently be the main indication of substandard fetal growth, this approach is prone to considerable errors. It has been suggested that FGR is undetected in about 30% of routinely scanned embryos, and incorrectly detected in 50% of cases [18]. Nonetheless, ultrasound is an invaluable tool for accurate pregnancy dating and diagnosis of FGR. Early in the first trimester of pregnancy, a crown-to-rump length that is within 5 days of the patient's menstrual dating serves to reliably confirm the gestational age. Prior to the visualization of an embryo, a gestational sac can be seen as early as 4.5 weeks using transvaginal ultrasound. The gestational sac grows at a mean diameter of 1 mm per day and can be used to determine the correct gestational age [19, 20]. In the second trimester and prior to 20 weeks of gestation, the fetal biometry within 10 days of the menstrual date is considered reliable. The accuracy of ultrasound to determine the correct gestational age is diminished beyond 20 weeks, when the biparietal diameter may diverge from the correct gestational age by 12–15 days. This discrepancy may extend to 21 days after 30–32 weeks of pregnancy, leading many researchers to utilize the femur length as a more dependable indicator of gestational age late in pregnancy [21, 22].

Sonographic estimation of fetal weight is achieved using polynomial equations combining the biparietal diameter, femur length and the abdominal circumference. The most commonly employed formulas are those reported by Shepard et al. [23] and Hadlock et al. [24]. Using the estimated fetal weight derived from these formulas, FGR is typically defined as weight less than the 10th, 5th or 3rd percentile for the gestational age or below 2 standard deviations of the mean for the gestational age [25, 26], possibly with adjustment to maternal weight, height, parity and ethnic background. The association of substandard growth of the abdominal circumference with normal growth of the head circumference has been suggested as a means of distinguishing asymmetrical growth restriction (suggesting placental dysfunction) from symmetrical growth restriction (constitutional smallness). This is attributed to the preferential shunting of blood to vital organs (brain, heart and adrenal glands) in cases of utero-placental hypoperfusion. Unfortunately, the etiology and manifestation of symmetric and asymmetric FGR largely overlap, limiting the usefulness of this classification in defining FGR and its clinical consequences [25]. Based on the assumption that fetal weight is proportional to fetal volume, attempts have been made to estimate the volume of the fetal body using 2-D and 3-D imaging technology [27, 28]. Because of the complexity of these measurements, they are currently not recommended for routine clinical practice, except for research in this area.

Ultrasonography may also shed light on placental size, morphology and amniotic fluid volume, which can be estimated using either subjective assessment or calculation of the amniotic fluid index (AFI) [29]. Because of the relative ease of AFI measurement it has become a useful tool in the initial analysis of FGR. For example, normal AFI may support normal placental perfusion, and suggest other etiologies of FGR such as infection with CMV, where AFI may be normal or increased. In contrast, low AFI with intact membranes is most commonly associated with utero-placental insufficiency or fetal anomalies.

Antepartum Fetal Monitoring

The goal of antepartum fetal monitoring in the setting of FGR is to detect fetuses at risk for acidosis-hypoxemia and determine the optimal timing for delivery before further deterioration that may lead to permanent fetal injury or stillbirth. Reassuring tests will delay the need for delivery, thereby diminishing the risk of iatrogenic prematurity. Frequently used monitoring tools include the nonstress test or cardiotocography and biophysical profile scoring, Doppler interrogation of different fetal blood vessels and fetal blood sampling. Additionally, maternal perception of fetal movements is an important clinical measure that requires no special technology. Fetal kick counts are helpful in monitoring the well-being of fetuses at risk for hypoxia, particularly in the outpatient setting.

Nonstress and Contraction Stress Test
The nonstress test (NST) or cardiotocogram (CTG) is a frequently used antenatal surveillance tool. The NST determines fetal heart rate baseline, variability and periodic changes. In a healthy fetus a normal baseline, the presence of variability and the presence of fetal heart rate accelerations (which are associated with fetal movements) reflect adequate oxygenation of the fetal central nervous system. A reactive NST after 32 weeks is defined by the presence of at least two sustained fetal heart rate accelerations of at least 15 beats lasting for greater than 15 s each during a 20-min monitoring period. The NST can be interpreted using visual inspection, or by computer assisted analysis. The computerized analysis of the NST may reduce some of the inconsistencies in interpretation, yet its advantage over standard interpretation remains to be proven. While NST is a well-accepted tool that is relatively easy to perform, its advantages are overshadowed by a very high false-positive rate for abnormalities, and a relatively high interobserver and intraobserver variation [30]. These limitations should be borne in mind when using NST for monitoring growth-restricted pregnancies. While commonly used as a first screening strategy, additional tools are available to refine the diagnostic accuracy of fetal well-being, as discussed below. Finally, a contraction stress test can be used as a more sensitive tool to uncover a dysfunctional placenta. During this test, the infusion of oxytocin causes uterine contractions, which are associated

with diminished placental perfusion pressure. Abnormalities in the fetal heart rate may be uncovered even prior to their appearance using a non-stress test, and identify a fetus at risk for hypo-oxygenation secondary to placental dysfunction.

Biophysical Profile

The fetal biophysical profile (BPP) is a group of measurements that includes the amniotic fluid volume, fetal tone, fetal movements, fetal breathing movements and fetal heart rate monitoring (NST). When normal, each parameter receives two points, for a maximum total of ten points. It is the most acceptable method of noncontinuous fetal well-being assessment in the USA [31–33]. The different components of the BPP reflect fetal functions that are indicative of fetal well-being. Whereas most components of the BPP serve as immediate markers of fetal status, the measurement of amniotic fluid volume serves mainly as an indirect estimate of fetal urine production, a marker of fetal renal perfusion. Although equal weight is given to each of the BPP parameters, the finding of oligohydramnios usually calls for further evaluation, irrespective of the overall score. When all other parameters are within normal limits, the need for an NST is questionable. In a high-risk pregnancy protocol described by Manning et al. [34], a routine NST was not performed when all other BPP parameters were normal. In contrast, others view the NST and the BPP as independent predictors of normal outcome [35]. While the BPP is usually employed to lower the false-positive rate of the NST, only a few studies addressed the false-positive rate of the BPP itself, which may range from 75% for a score of six to 20% for a score of zero. The use of a vibro-acoustic stimulator, designed to stimulate fetal activity during the BPP test, has been suggested as a means to reduce the false-positive rate [36]. Notably, although the main advantages of the BPP test are the direct assessment of fetal behavior and the relative ease of performance (when an ultrasound scanner is available), this test requires approximately 30 min of a technologist's time, depends on visual interpretation of the NST, and provides only indirect information on fetal cardiovascular status and perfusion. Although the BPP is popular in North America, randomized trials comparing it with other testing modalities are lacking [37].

Doppler Velocimetry of Blood Flow

This technology focuses on the determination of blood flow in fetal vessels as a means to define vascular resistance and organ function. Although Doppler velocimetry of the uterine artery has been used to interrogate uterine perfusion and thereby define those at risk for FGR or other complications of pregnancy [38], this measure focuses on maternal perfusion abnormalities, and not on conditions inherent to the placenta. The role of uterine artery flow analysis is therefore limited, particularly after the diagnosis of FGR has been established. The fetal blood vessels that have been most intensely studied using Doppler flow techniques include the umbilical artery (UA; fig. 1, upper panel), which evaluates blood flow from the fetus to the placenta and therefore abnormalities in placental vessel resistance, the middle cerebral artery (MCA; fig. 1, middle panel), which

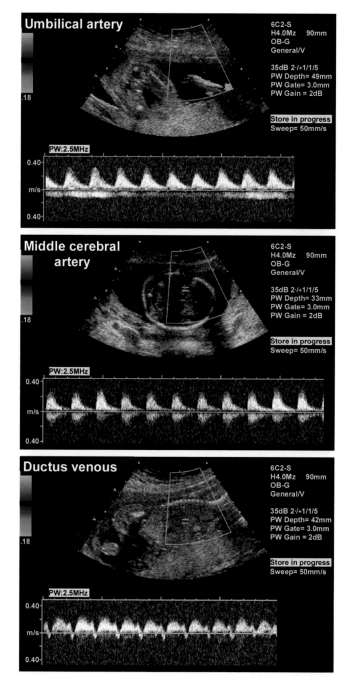

Fig. 1. Doppler velocimetry of fetal vessels. Upper panel: normal umbilical artery flow waveform. Middle panel: normal waveform of the middle cerebral artery. Lower panel: abnormal ductus venosus waveform, showing a reversed 'a' wave.

interrogates fetal brain blood flow, and the ductus venosus (DV; fig. 1, lower panel), where alterations in flow represent diminished cardiac function in more severe cases of FGR [39, 40]. The UA is the most commonly used Doppler surveillance tool in women diagnosed with FGR. An abnormal UA flow points to placental dysfunction, and is

Odibo · Sadovsky

therefore commonly used to differentiate fetal growth abnormality that reflects FGR from those reflecting SGA. In the early stages of placental dysfunction there is an increase in the resistance to blood flow through the UA, manifesting as increased systolic/diastolic (S/D) flow ratio (also measured as pulsatility index, defined as systolic minus diastolic flow, divided by the mean). With worsening placental insufficiency, diastolic flow becomes absent or reversed, which entails a greater risk of hypoxia, with increased perinatal mortality and morbidity [41–43]. The use of UA Doppler as a part of the monitoring scheme for pregnancies complicated by FGR may reduce mortality rate (OR 0.67, 95% CI, 0.47–0.97) and also lower the need for interventions such as antepartum admissions, labor induction and cesarean deliveries [43, 44].

The rationale for Doppler flow assessment of the MCA is based on blood flow shunting from visceral, less-essential organs, to vital fetal organs such as the brain, heart and adrenal glands. Autoregulatory mechanisms, known as 'brain-sparing' effect, enhance the blood flow to the brain, resulting in decreased systolic/diastolic ratios (fig. 1, middle panel). Notably, profound hypoperfusion or fetal hypoxia may surpass this defense mechanism, leading to a rebound increase in S/D ratio and diminished perfusion even to the brain. Unfortunately, these changes were not proven to reliably predict fetal worsening, or assist in the management of pregnancies complicated by FGR [45, 46]. It remains to be determined whether or not the combination of elevated MCA pulsatility indices with UA Doppler may have a role in optimizing the timing of the delivery of the growth restricted fetus [44, 47, 48]. The determination of MCA peak systolic velocity (as presently used to diagnose and manage fetal anemia) may assist in the follow up of fetuses with an established diagnosis of FGR [49].

Doppler analyses of fetal veins that are currently evaluated for clinical practice include the DV, inferior and superior vena cava and the umbilical vein. Changes in blood flow in these vessels reflect the compliance of the fetal heart and its ability to handle the preload volume. For example a decreased or reversed 'a' wave of the DV (fig. 1, lower panel) is evidence of decreased forward flow in atrial systole, while pulsations in the umbilical vein reflect increased central venous pressure [50]. At present, it is not clear whether or not there is a reproducible 'sequence' of venous flow alterations in the growth restricted fetus. Several of the changes in blood flow, particularly in the umbilical vein and ductus venosus, seem to temporally overlap. Whereas arterial and venous Doppler indices may independently predict fetal deterioration, further studies are needed to determine (a) the sequence of appearance of abnormal Doppler parameters in growth restricted fetuses with placental dysfunction, and (b) the possible integration of different tests for better prediction of outcome and the timing of intervention [50–52].

Cordocentesis
Whereas fetal blood sampling, designed to assess fetal pH, pCO_2 and bicarbonate level, might be considered the most direct indicators of fetal well-being, the invasive

nature of this procedure limits its role in monitoring the status of the growth-restricted fetus. Even when performed, cordocentesis provides a snapshot analysis at a single time-point instead of the temporal changes that can be easily assessed by the indirect modalities discussed above.

Histopathological and Molecular Diagnostics

Because the placenta plays a pivotal role in supporting growth and development of the mammalian fetus, it is not surprising that placental malfunction frequently underlies the development of FGR. Branching fetal vessels in normal placental villi are supported by connective tissue. This connective tissue is lined by a trophoblast bilayer that is bathed in maternal blood. Multinucleated syncytiotrophoblasts and mononuclear cytotrophoblasts within the villous bilayer regulate gas exchange, nutrient transport to the fetus, removal of waste to the maternal blood, placental hormone production and immune defense. The histological changes that characterize placental injury in pregnancies complicated by growth-restricted fetuses are relatively nonspecific, and include damage to branching angiogenesis with long unbranched intermediate and terminal villi, cytotrophoblast proliferation, trophoblast apoptosis, fibrin deposition, syncytial knotting and bridging and enhanced villous maturation [14, 53].

Recent advances in the use of molecular probes for high-throughput analysis of clinical conditions has advanced our ability to comprehensively interrogate disease mechanisms and further refine the laboratory-based assessment of placental injury. Genomic, proteomics and metabolomic tools can also usher the way to improvement of disease diagnosis, and generate new means for therapeutic interventions, designed to remove a cause of a disease or ameliorate the undesired outcome. The use of such tools in the clinical approach to FGR depends on acquisition of tissues or cells from the feto-placental units. Although these approaches are still in their infancy in the context of FGR, feto-placental tissues are potentially accessible to researchers and clinicians through sampling of amniotic fluid, fetal blood, maternal blood and trans-cervical or trans-abdominal access to placenta tissues. We have recently demonstrated that placental villi obtained after delivery from human pregnancies complicated by FGR exhibit a characteristic set of changes in 'hypoxic trophoblast signature transcripts' [54]. Using high-density oligonucleotide microarrays we initially examined differences in gene expression between trophoblast cultured in standard or hypoxic conditions ($\leq$1% oxygen), as well as in placental tissues from pregnancies complicated by FGR vs. matched controls. Among 44 recognized gene products that exhibited the most dramatic (up or down) difference we used in situ hybridization and quantitative PCR to confirm the upregulation of transcripts for vascular endothelial growth factor, connective tissue growth factor, follistatin-related protein, N-Myc downstream-regulated gene1 and adipophilin (ADRP), and the downregulation of human placental lactogen and PHLDA2 [54]. Others have found additional transcripts to be dysregulated in villi from pregnancies complicated by FGR, including an upregulation of CRH, HPGD,

INHB4, LEP and PSG4, and downregulation of IGF1, IGF2, AGTR1, DSCR1 and GATM. These and other gene products (such as leptin and sFlt) were confirmed by similar studies using human placentas. Interestingly, several imprinted genes, such as the maternally expressed/paternally repressed *Phlda2* or the paternally expressed/maternally repressed gene *Mest* are differentially expressed in placentas from FGR fetuses [55]. This is particularly intriguing, because these genes are known regulators of feto-placental growth, with paternally expressed genes enhancing growth and maternally expressed genes diminishing growth [56]. Taken together, it is likely that these hypoxic trophoblast signature transcripts may implicate mediators in pathways underlying trophoblast hypoxic injury and adaptation, and also serve as future diagnostic tools for FGR associated with placental injury. It is also plausible that high-throughput proteomic methods, designed to elucidate biological pathways underlying complex diseases, may complement the genomic approaches described earlier. Indeed, several protein families, such cytokines, growth factors and angiogenic peptides have been implicated in the pathogenesis of FGR [57]. The integration of fetal biophysical testing, combined with informatics-based molecular analysis may advance our understanding of disease pathophysiology and suggest new biomarkers for early and reliable diagnosis of FGR.

Clinical Management of FGR

Therapeutic Interventions to Abolish or Attenuate the Impact of FGR
Although diverse interventions have been attempted in order to improve the outcome or mitigate the insult that leads to FGR, most interventions have generally had little impact on the clinical outcome or its sequelae [58]. It is clear that attempts should be made to remove the cause of feto-placental injury, such as improved nutrition, smoking cessation, treatment of infectious diseases, avoidance of illicit drugs and control of maternal disorders including hypertension and renal dysfunction. It is also imperative to exclude any lethal fetal malformations that could make any further management irrelevant. Therefore, ultrasonographic evaluation for such malformations and options of fetal karyotyping must be considered. Specific interventions such as plasma volume expansion, oxygen supplementation, administration of glucose or amino acids and administration of low-dose aspirin to the mother have generally resulted in minimal or no impact on FGR [58–60]. In a meta-analysis of more than 120 studies evaluating interventions aimed at prevention or treatment of FGR, Gulmezoglu et al. [58] reported that most of the interventions did not have a significant impact on perinatal outcomes. Whereas smoking cessation and the use of antimalarial therapy appeared to prevent FGR, these interventions are ineffective once FGR is already established. Notably, a recent Cochrane review suggests that nutritional advice to women and balanced energy/protein supplements may be beneficial in ameliorating the disease risk [61, 62].

Timing the Delivery of the Growth-Restricted Fetus

In the absence of interventions that clearly improve fetal growth, most investigations have focused on the use of biophysical or biochemical tools designed to optimize the timing of delivery. If FGR is diagnosed at full term ($\geq$37 weeks by reliable dates), the decision to proceed with delivery is relatively straightforward. Even in pregnancies at 34–37 weeks, when the incidence of the most significant neonatal complications is rather low and the risk of hyaline membrane disease can be assessed by amniocentesis, the decision is less problematic [63]. The decision-making process becomes more complex with earlier gestational ages (see below). Existing studies regarding the benefits of antepartum surveillance for pregnancies complicated by FGR are limited by study design and inclusion criteria. Most studies did not solely target FGR, but included other high-risk conditions such as preeclampsia or diabetes. In a recent review of four studies involving 1,588 high or intermediate risk pregnancies, antenatal NST appeared to have no significant effect on perinatal mortality or morbidity [64]. Similarly, a review by the Cochrane researchers on the role of BPP in the management of high-risk pregnancies found no difference between biophysical profile and other forms of fetal assessment [37]. In contrast, studies by Manning et al. [32] suggest that NST or BPP may define high-risk pregnancies where the risk of in utero demise is high, and therefore early delivery might be warranted. Recent evidence suggests that the use of UA Doppler may reduce the need for antepartum interventions and lower perinatal morbidity and mortality [43]. Unlike UA Doppler, the use of this technique to interrogate flow and vessel resistance in other vascular beds, such as MCA, ductus venosus or other venous vessels has not been rigorously studied.

Currently, there is no single test that indicates the optimal timing of delivery. Considerations for delaying the delivery as much as feasible in order allow the fetus to gain maturity are counter-balanced not only by the need to deliver prior to fetal death, but also prior to any permanent injury to the brain or other vital fetal organs. The European Growth Restriction Intervention Trial (GRIT) [65], one of the only randomized trials in this field, was designed to compare the effect of preterm delivery (at 24–36 weeks), based on UA Doppler waveform, with that of delayed delivery by other clinical indicators. Women were randomly assigned to immediate delivery for abnormal UA Doppler velocimetry or a delayed delivery until the managing physician believed that delivery was warranted, based on worsening tests or a favorable gestational age. The main outcome variables were survival to hospital discharge and developmental quotient at two years of age. Of 548 pregnancies randomized into the study, there was no difference in overall mortality between the immediate delivery and the delayed delivery groups [65]. In the 2-year follow-up study, there were no significant differences between the groups in death or disability rates [66]. Interestingly, the fact that the managing physicians were prepared to time the delivery even as early as 29–34 weeks using a randomization process points to the uncertainty regarding risk/benefit of immediate delivery. Unfortunately, the

GRIT study did not address specific triggers for timing of delivery. Therefore, the question of the best indicator for delivery remains unanswered. Using published data we developed a decision tree to explore the optimal antepartum test for timing the delivery of the preterm FGR fetus [67]. Although our retrospective decision analysis indicated that BPP was the best test to guide decisions on delivery of the preterm growth-restricted fetus, it is clear that our results must be corroborated by a well-designed, prospective clinical trial prior to universal acceptance. In the absence of a single test, different centers currently depend on either biophysical tests (NST, computerized NST, BPP and alike) or Doppler blood flow studies in order to optimize the timing of delivery. Figure 2 depicts a suggested scheme for antenatal monitoring of the growth-restricted fetus. This scheme is not prescriptive, and should be considered as a proposed pathway. The decision to deliver a growth-restricted fetus must be individualized and commensurate with prevailing local neonatal facilities. Notably, in most centers where perinatal-neonatal services are available, pregnancies complicated by FGR frequently do not proceed to term and beyond. After 34 completed weeks the appearance of advanced, worsening signs of fetal deterioration, such as absent or reversed UA diastolic flow, persistent non-reassuring NST, a BPP score of ≤4, reversed 'a' wave of the ductus venosus or umbilical vein pulsations may suggest the need for immediate delivery [68]. It is also clear that administration of a course of steroids, which has been repeatedly shown to reduce the incidence of hyaline membrane disease and perhaps mitigate other neonatal complications, should be considered in every preterm, growth-restricted fetus.

Mode of Delivery of the Growth-Restricted Fetus
Although only a few studies have addressed the optimal mode of delivery of the growth-restricted fetus, it seems prudent that a vaginal delivery should not be attempted when obvious biophysical signs of nonreassuring fetal status are present prior to uterine contractions. Because placental perfusion is virtually halted during a contraction, fetal stress is likely to be exacerbated during labor. The mode of delivery for the growth-restricted fetus that exhibits reassuring biophysical parameters is presently uncertain. In the GRIT study, one third of the pregnancies with FGR were delivered by the abdominal route, yet it is not clear if there was any benefit from this approach [65]. A recent Cochrane review pointed out that cesarean delivery for SGA fetuses was associated with a lower rate of respiratory distress syndrome, neonatal seizures and death, but these trends were statistically insignificant [69]. Obviously, other obstetrical factors such as the gestational age, cervical status, fetal presentation and maternal medical complications may influence the choice of delivery route. Because the growth-restricted fetus may require specialized neonatal care, particularly when the delivery occurs preterm, it is prudent to transfer the fetus to a well-equipped center where experienced perinatal-neonatal care is available.

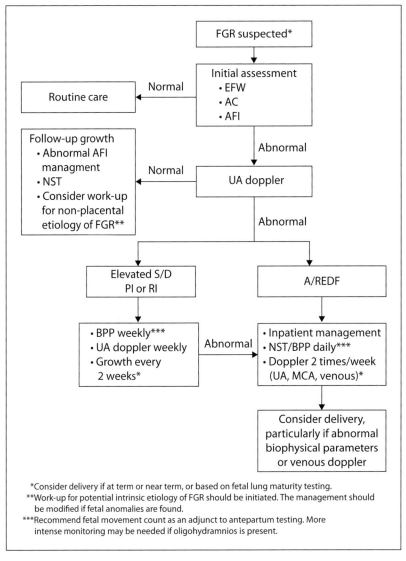

Fig. 2. A proposed monitoring scheme in pregnancies complicated by FGR. This figure focuses on FGR associated with placental dysfunction. See text, for additional details.

Risk of Recurrence of FGR

In general, having a pregnancy complicated by FGR increases the risk of recurrence of this condition [7, 16], yet this risk depends on the etiology of the FGR. For example, when FGR is attributed to an underlying chronic maternal medical condition, the likelihood of recurrence is high. In contrast, FGR caused by intrauterine infection is less likely to recur. It is therefore paramount to attempt to identify the causative factor in order to assist with neonatal care, and adequately counsel the parents regarding the likelihood of FGR in future pregnancies.

Conclusion

FGR is a complex, multifactorial disease that frequently complicates pregnancy. Although heterogeneous etiologies may contribute to growth-restriction, identification of the causes may assist the clinician in optimizing management decisions, including the timing of delivery. Unfortunately, most commonly used tools to assess fetal status are indirect, and their predictive value is limited. Furthermore, the temporal correlation between the commonly used biophysical tests of fetal well being and the deterioration of fetal status is currently unclear. This uncertainty, combined with fact that most management strategies for pregnancies complicated by FGR are based on limited (level II–III) epidemiologic evidence, underscores the need for well-designed randomized clinical trials that target different management options.

References

1   World Health Organization: International Statistical Classification of Diseases and Related Health Problems. Geneva, World Health Organization, 1992.

2   Doubilet PM, Benson CB: Ultrasound evaluation of fetal growth; in Callen PW (ed): Ultrasonography in Obstetrics and Gynecology. Philadelphia, Saunders, 2000.

3   McIntire DD, Bloom SL, Casey BM, Leveno KJ: Birth weight in relation to morbidity and mortality among newborn infants. N Engl J Med 1999;340: 1234–1238.

4   Bernstein I, Gabbe SG, Reed KL: Intrauterine growth restriction; in Gabbe SG, Niebyl JR, Simpson JL (eds): Obstetrics: Normal and Problem Pregnancies. New York, Churchill-Livingstone, 2002.

5   Soothill PW, Nicolaides KH, Campbell S: Prenatal asphyxia, hyperlacticaemia, hypoglycaemia, and erythroblastosis in growth retarded fetuses. Br Med J (Clin Res Ed) 1987;294:1051–1053.

6   Barker DJ, Winter PD, Osmond C, Margetts B, Simmonds SJ: Weight in infancy and death from ischaemic heart disease. Lancet 1989;ii:577–580.

7   Ounsted M, Moar VA, Scott A: Risk factors associated with small-for-dates and large-for-dates infants. Br J Obstet Gynaecol 1985;92:226–232.

8   Cunningham FG, Cox SM, Harstad TW, Mason RA, Pritchard JA: Chronic renal disease and pregnancy outcome. Am J Obstet Gynecol 1990;163:453–459.

9   Kupferminc MJ, Peri H, Zwang E, Yaron Y, Wolman I, Eldor A: High prevalence of the prothrombin gene mutation in women with intrauterine growth retardation, abruptio placentae and second trimester loss. Acta Obstet Gynecol Scand 2000;79: 963–967.

10   Naeye RL: Prenatal organ and cellular growth with various chromosomal disorders. Biol Neonate 1967; 11:248–260.

11   Khoury MJ, Erickson JD, Cordero JF, McCarthy BJ: Congenital malformations and intrauterine growth retardation: a population study. Pediatrics 1988;82: 83–90.

12   Snijders RJ, Sherrod C, Gosden CM, Nicolaides KH: Fetal growth retardation: associated malformations and chromosomal abnormalities. Am J Obstet Gynecol 1993;168:547–555.

13   Fowler KB, Stagno S, Pass RF, Britt WJ, Boll TJ, Alford CA: The outcome of congenital cytomegalovirus infection in relation to maternal antibody status. N Engl J Med 1992;326:663–667.

14   Kingdom JC, Kaufmann P: Oxygen and placental villous development: origins of fetal hypoxia. Placenta 1997;18:613–621, discussion 623–616.

15   Lestou VS, Kalousek DK: Confined placental mosaicism and intrauterine fetal growth. Arch Dis Child Fetal Neonatal Ed 1998;79:F223–F226.

16   Alberry M, Soothill P: Management of fetal growth restriction. Arch Dis Child Fetal Neonatal Ed 2007; 92:F62–F67.

17   Bailey SM, Sarmandal P, Grant JM: A comparison of three methods of assessing inter-observer variation applied to measurement of the symphysis-fundal height. Br J Obstet Gynaecol 1989;96:1266–1271.

18   Jahn A, Razum O, Berle P: Routine screening for intrauterine growth retardation in Germany: low sensitivity and questionable benefit for diagnosed cases. Acta Obstet Gynecol Scand 1998;77:643–648.

19 Nyberg DA, Mack LA, Laing FC, Patten RM: Distinguishing normal from abnormal gestational sac growth in early pregnancy. J Ultrasound Med 1987; 6:23–27.

20 Robinson HP, Fleming JE: A critical evaluation of sonar 'crown-rump length' measurements. Br J Obstet Gynaecol 1975;82:702–710.

21 Hadlock FP, Deter RL, Harrist RB, Park SK: Fetal biparietal diameter: a critical re-evaluation of the relation to menstrual age by means of real-time ultrasound. J Ultrasound Med 1982;1:97–104.

22 Benson CB, Doubilet PM: Sonographic prediction of gestational age: accuracy of second- and third-trimester fetal measurements. Am J Roentgenol 1991;157:1275–1277.

23 Shepard MJ, Richards VA, Berkowitz RL, Warsof SL, Hobbins JC: An evaluation of two equations for predicting fetal weight by ultrasound. Am J Obstet Gynecol 1982;142:47–54.

24 Hadlock FP, Harrist RB, Sharman RS, Deter RL, Park SK: Estimation of fetal weight with the use of head, body, and femur measurements: a prospective study. Am J Obstet Gynecol 1985;151:333–337.

25 Blackwell SC, Moldenhauer J, Redman M, Hassan SS, Wolfe HM, Berry SM: Relationship between the sonographic pattern of intrauterine growth restriction and acid-base status at the time of cordocentensis. Arch Gynecol Obstet 2001;264:191–193.

26 Gardosi J, Chang A, Kalyan B, Sahota D, Symonds EM: Customised antenatal growth charts. Lancet 1992;339:283–287.

27 Combs CA, Jaekle RK, Rosenn B, Pope M, Miodovnik M, Siddiqi TA: Sonographic estimation of fetal weight based on a model of fetal volume. Obstet Gynecol 1993;82:365–370.

28 Schild RL, Fimmers R, Hansmann M: Fetal weight estimation by three-dimensional ultrasound. Ultrasound Obstet Gynecol 2000;16:445–452.

29 Moore TR, Cayle JE: The amniotic fluid index in normal human pregnancy. Am J Obstet Gynecol 1990;162:1168–1173.

30 Cheng LC, Gibb DM, Ajayi RA, Soothill PW: A comparison between computerised (mean range) and clinical visual cardiotocographic assessment. Br J Obstet Gynaecol 1992;99:817–820.

31 Manning FA, Morrison I, Harman CR, Menticoglou SM: The abnormal fetal biophysical profile score. V. Predictive accuracy according to score composition. Am J Obstet Gynecol 1990;162:918–924, discussion 924–917.

32 Manning FA, Harman CR, Morrison I, Menticoglou SM, Lange IR, Johnson JM: Fetal assessment based on fetal biophysical profile scoring. IV. An analysis of perinatal morbidity and mortality. Am J Obstet Gynecol 1990;162:703–709.

33 Vintzileos AM, Gaffney SE, Salinger LM, Kontopoulos VG, Campbell WA, Nochimson DJ: The relationships among the fetal biophysical profile, umbilical cord pH, and Apgar scores. Am J Obstet Gynecol 1987;157:627–631.

34 Manning FA, Morrison I, Lange IR, Harman CR, Chamberlain PF: Fetal biophysical profile scoring: selective use of the nonstress test. Am J Obstet Gynecol 1987;156:709–712.

35 Platt LD, Eglinton GS, Sipos L, Broussard PM, Paul RH: Further experience with the fetal biophysical profile. Obstet Gynecol 1983;61:480–485.

36 Inglis SR, Druzin ML, Wagner WE, Kogut E: The use of vibroacoustic stimulation during the abnormal or equivocal biophysical profile. Obstet Gynecol 1993;82:371–374.

37 Alfirevic Z, Neilson JP: Biophysical profile for fetal assessment in high risk pregnancies. Cochrane Database Syst Rev 2000:CD000038.

38 Albaiges G, Missfelder-Lobos H, Lees C, Parra M, Nicolaides KH: One-stage screening for pregnancy complications by color Doppler assessment of the uterine arteries at 23 weeks' gestation. Obstet Gynecol 2000;96:559–564.

39 Baschat AA, Gembruch U, Harman CR: The sequence of changes in Doppler and biophysical parameters as severe fetal growth restriction worsens. Ultrasound Obstet Gynecol 2001;18:571–577.

40 Baschat AA: Fetal responses to placental insufficiency: an update. BJOG 2004;111:1031–1041.

41 Weiner CP: The relationship between the umbilical artery systolic/diastolic ratio and umbilical blood gas measurements in specimens obtained by cordocentesis. Am J Obstet Gynecol 1990;162:1198–1202.

42 Kingdom JC, Burrell SJ, Kaufmann P: Pathology and clinical implications of abnormal umbilical artery Doppler waveforms. Ultrasound Obstet Gynecol 1997;9:271–286.

43 Alfirevic Z, Neilson JP: Doppler ultrasonography in high-risk pregnancies: systematic review with meta-analysis. Am J Obstet Gynecol 1995;172:1379–1387.

44 Westergaard HB, Langhoff-Roos J, Lingman G, Marsal K, Kreiner S: A critical appraisal of the use of umbilical artery Doppler ultrasound in high-risk pregnancies: use of meta-analyses in evidence-based obstetrics. Ultrasound Obstet Gynecol 2001; 17:466–476.

45 Ott WJ: Intrauterine growth restriction and Doppler ultrasonography. J Ultrasound Med 2000;19: 661–665.

46 Sterne G, Shields LE, Dubinsky TJ: Abnormal fetal cerebral and umbilical Doppler measurements in fetuses with intrauterine growth restriction predicts the severity of perinatal morbidity. J Clin Ultrasound 2001;29:146–151.

47 Madazli R, Uludag S, Ocak V: Doppler assessment of umbilical artery, thoracic aorta and middle cerebral artery in the management of pregnancies with growth restriction. Acta Obstet Gynecol Scand 2001; 80:702–707.

48 Odibo AO, Riddick C, Pare E, Stamilio DM, Macones GA: Cerebroplacental Doppler ratio and adverse perinatal outcomes in intrauterine growth restriction: evaluating the impact of using gestational age-specific reference values. J Ultrasound Med 2005;24:1223–1228.

49 Mari G, Hanif F, Kruger M, Cosmi E, Santolaya-Forgas J, Treadwell MC: Middle cerebral artery peak systolic velocity: a new Doppler parameter in the assessment of growth-restricted fetuses. Ultrasound Obstet Gynecol 2007;29:310–316.

50 Baschat AA, Gembruch U, Reiss I, Gortner L, Weiner CP, Harman CR: Relationship between arterial and venous Doppler and perinatal outcome in fetal growth restriction. Ultrasound Obstet Gynecol 2000;16:407–413.

51 Senat MV, Schwarzler P, Alcais A, Ville Y: Longitudinal changes in the ductus venosus, cerebral transverse sinus and cardiotocogram in fetal growth restriction. Ultrasound Obstet Gynecol 2000;16: 19–24.

52 Baschat AA, Galan HL, Bhide A, Berg C, Kush ML, Oepkes D, Thilaganathan B, Gembruch U, Harman CR: Doppler and biophysical assessment in growth restricted fetuses: distribution of test results. Ultrasound Obstet Gynecol 2006;27:41–47.

53 Benirschke K, Kaufmann P: Pathology of the Human Placenta. New York, Springer, 2000.

54 Roh CR, Budhraja V, Kim HS, Nelson DM, Sadovsky Y: Microarray-based identification of differentially expressed genes in hypoxic term human trophoblasts and in placental villi of pregnancies with growth restricted fetuses. Placenta 2005;26:319–328.

55 McMinn J, Wei M, Sadovsky Y, Thaker HM, Tycko B: Imprinting of PEG1/MEST isoform 2 in human placenta. Placenta 2006;27:119–126.

56 Tycko B, Morison IM: Physiological functions of imprinted genes. J Cell Physiol 2002;192:245–258.

57 Shankar R, Cullinane F, Brennecke SP, Moses EK: Applications of proteomic methodologies to human pregnancy research: a growing gestation approaching delivery? Proteomics 2004;4:1909–1917.

58 Gulmezoglu M, de Onis M, Villar J: Effectiveness of interventions to prevent or treat impaired fetal growth. Obstet Gynecol Surv 1997;52:139–149.

59 Nicolaides KH, Campbell S, Bradley RJ, Bilardo CM, Soothill PW, Gibb D: Maternal oxygen therapy for intrauterine growth retardation. Lancet 1987;i: 942–945.

60 Nicolini U, Hubinont C, Santolaya J, Fisk NM, Rodeck CH: Effects of fetal intravenous glucose challenge in normal and growth retarded fetuses. Horm Metab Res 1990;22:426–430.

61 Kramer MS, Kakuma R: Energy and protein intake in pregnancy. Cochrane Database Syst Rev 2003: CD000032.

62 Fawzi WW, Msamanga GI, Urassa W, Hertzmark E, Petraro P, Willett WC, Spiegelman D: Vitamins and perinatal outcomes among HIV-negative women in Tanzania. N Engl J Med 2007;356:1423–1431.

63 Sinclair JC: Meta-analysis of randomized controlled trials of antenatal corticosteroid for the prevention of respiratory distress syndrome (discussion). Am J Obstet Gynecol 1995;173:335–344.

64 Pattison N, McCowan L: Cardiotocography for antepartum fetal assessment. Cochrane Database Syst Rev 2000:CD001068.

65 GRIT Study Group: A randomised trial of timed delivery for the compromised preterm fetus: short term outcomes and Bayesian interpretation. BJOG 2003;110:27–32.

66 Thornton JG, Hornbuckle J, Vail A, Spiegelhalter DJ, Levene M: Infant wellbeing at 2 years of age in the Growth Restriction Intervention Trial (GRIT): multicentred randomised controlled trial. Lancet 2004;364:513–520.

67 Odibo AO, Quinones JN, Lawrence-Cleary K, Stamilio DM, Macones GA: What antepartum fetal test should guide the timing of delivery of the preterm growth-restricted fetus? A decision-analysis. Am J Obstet Gynecol 2004;191:1477–1482.

68 Maulik D: Management of fetal growth restriction: an evidence-based approach. Clin Obstet Gynecol 2006;49:320–334.

69 Grant A, Glazener CM: Elective caesarean section versus expectant management for delivery of the small baby. Cochrane Database Syst Rev 2001: CD000078.

Yoel Sadovsky, MD
Magee Womens Research Institute
204 Craft Avenue, Room A608
Pittsburgh, PA 15213 (USA)
Tel. +1 412 641 2675, Fax +1 412 641 3898, E-Mail ysadovsky@mwri.magee.edu

Kiess W, Chernausek SD, Hokken-Koelega ACS (eds): Small for Gestational Age. Causes and Consequences.
Pediatr Adolesc Med. Basel, Karger, 2009, vol 13, pp 26–43

Fetal Growth Restriction and the Developmental Origins of Adult Disease Hypothesis: Experimental Studies and Biological Consequences

Stefan O. Krechowec · Nichola M. Thompson ·
Bernhard H. Breier

Liggins Institute and National Research Centre for Growth and Development,
University of Auckland, Auckland, New Zealand

Abstract

The clinical term small for gestational age (SGA) is used to describe newborn infants whose birth weight falls below the 10th percentile of weight for their gestational age. This classification encompasses two broad developmental phenotypes, infants which have achieved their genetic potential for growth but are constitutionally small, and infants that have experienced fetal growth restriction (FGR) and have failed to achieve their relative growth potential. The diagnosis of FGR has an immediate clinical relevance for SGA babies because FGR is associated with an increased risk of perinatal complications, including preventable neonatal mortality. Due to the serious nature of such perinatal complications there has been a great deal of interest in identifying and defining the immediate causes and consequences of FGR. Currently FGR is commonly described as a consequence of adverse prenatal environmental stimuli that can impair gas exchange and nutrient delivery to the fetus disrupting the normal patterns of growth and development. Placental dysfunction and adverse maternal factors such as, smoking, maternal disease and malnutrition are considered key pathological factors responsible for FGR. While the majority of clinical research has focused on the immediate neonatal consequences of FGR, recent research over the past 15 years suggests that FGR can adversely affect postnatal health outcomes well into adulthood. This concept is called the developmental origins of health and disease hypothesis (DOHaD).

Developmental Origins of Health and Disease Hypothesis

Formerly known as the Barker [1] or fetal origins hypothesis, the developmental origins of health and disease (DOHaD) hypothesis proposes that the risks of developing chronic disease in adulthood are determined by adverse early life experiences (e.g. fetal

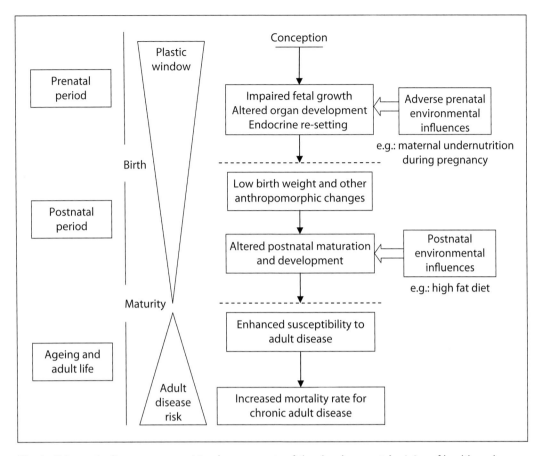

Fig. 1. Schematic diagram summarising key concepts of the developmental origins of health and disease hypothesis.

growth restriction; FGR) which alter normal patterns of growth and development. More specifically, it is hypothesised that adverse prenatal environmental stimuli acting during sensitive periods of fetal development induce permanent changes in growth and development that in later adult life increase an individuals susceptibility to adult disease [1] (fig. 1). The theoretical foundations of the DOHaD hypothesis rest upon the early epidemiological studies of Professor David Barker and colleagues. These studies found that indirect anthropomorphic markers of FGR, such as low birth weight, reduced head and abdominal circumference and low ponderal index were associated with an increased incidence of CHD [2, 3], hypertension [2, 4], and type II diabetes [5, 6] in adulthood. Notably, these relationships between birth size and adult disease risk extended across the normal range of birth weights and were not confined to clinically undersized babies [2, 5]. Barker's original hypothesis has proved to be highly controversial, sparking sharp debate over the existence of a relationship between impaired fetal growth and adult disease risk. Criticism has focused on the methodological approaches used in the original epidemiological studies [7, 8], on the relative importance of FGR and maternal undernutrition as causal mechanisms [9, 10], and on the strength of the relationship within the general population [11, 12]. Despite the vigorous debate, Barker's hypothesis has survived and evolved into the present DOHaD hypothesis which recognises a broad spectrum of early-life influences and developmental patterns that may be involved in determining adult disease risk, independently of clinically apparent FGR. Over the past 15 years, a sizeable body of clinical and experimental evidence has been generated which supports the fundamental biological plausibility of the DOHaD hypothesis. Scientific debate has shifted away from doubt over the existence of a link between early life environment and future disease risk and now focuses on the public health implications and causal mechanisms which underlie this relationship.

Animal Studies Testing the Developmental Origins Hypothesis

The DOHaD hypothesis describes a complex interaction, between early developmental events and adult disease processes, which evolves over the life course of an individual. The inherent costs and complexities of studying this phenomenon within a clinical setting have made animal studies a main tool of researchers investigating the DOHaD hypothesis. Many different animal models have been used to test the DOHaD hypothesis. However, because of its relatively short gestation period and quick postnatal maturation a majority of investigators have favoured the rat model. The following sections focus on the different experimental approaches that have been developed in the rat. In general, these models can be divided into three different categories on the basis of the type of environmental stimulus used to influence early development. These categories include nutritional, hormonal, and functional interventions.

Nutritional Interventions during Early Development

Altering fetal nutrition, in the rat, during pregnancy, is the most commonly used experimental approach to influence prenatal development. Maternal undernutrition during gestation was originally highlighted by Barker [1] as a primary causal mechanism and consequently this intervention was one of the first to be developed in animal studies. Two different approaches have been used to induce maternal undernutrition during pregnancy. The first approach restricts total maternal food intake, applying a global restriction of a nutritionally balanced diet during pregnancy [13] and/or lactation [14]. This manipulation is well known to impair both fetal and neonatal growth in the rat resulting in FGR [15]. The second approach restricts specific macro-nutrient components of the maternal diet causing an imbalanced undernutrition during pregnancy; the most frequently utilized is maternal protein restriction [16].

Maternal Undernutrition Models

A variety of experimental rat studies have used global maternal nutritional restriction, differing in the magnitude, timing, and duration of dietary restriction. The levels of maternal food restriction that have been used include 70% [17], 50% [14] and 30% [13] of normal food intake during pregnancy. The timing of dietary restriction has been constrained to early [17] or late [18] gestation, or alternatively maintained throughout the duration of pregnancy [13]. Results from these studies demonstrate that exposure to maternal undernutrition can have long-term outcomes on postnatal physiology. Exposure to 70% of normal maternal intake in the first half of pregnancy has been found to cause permanent elevations in blood pressure and increased responsiveness to vasoconstrictors within male offspring [17]. In contrast, a 50% restriction in the second half of gestation and during lactation leads to lasting impairments in glucose homeostasis [19], pancreatic development [14], lung maturation [20], and alterations to the hypothalamic-pituitary-adrenal (HPA) axis in both adult male and female offspring [21]. Similarly, an exposure to 30% of maternal food intake throughout gestation is associated with elevations in systolic blood pressure [13], reductions in kidney mass [13], hyperphagia [22], obesity [22, 23], and fasting hyperinsulinaemia in both adult male and female offspring [22, 23].

As a proof of principle, these experimental models demonstrate a plausible role for maternal undernutrition in the development of adult disease risk. However, the use of global maternal undernutrition has been criticised for a lack of applicability in today's modern society, where the majority of women have access to adequate levels of nutrition during pregnancy [10]. Nevertheless, maternal undernutrition still provides a useful model for many developing nations where significant maternal nutritional deficiencies still exist, and low birth weights, as a result, are common [24].

Low Protein Models

One of the most extensively studied rat models uses maternal protein restriction during pregnancy to influence fetal development. The maternal low protein (MLP) model involves feeding pregnant rat dams a diet containing 5–8% (weight/weight) protein (casein) during gestation, generally a little under half the protein content but equivalent in energy content [16, 25–27]. In general, offspring exposed to a MLP diet during pregnancy exhibit FGR with reduced birth weights and body length [26, 28, 29]. During lactation, catch-up in body weight can occur in the absence of a low protein diet, however, if MLP feeding is maintained during the weaning period postnatal growth can be permanently restricted [28]. A variety of adverse postnatal outcomes have been identified in offspring. Notably, these findings have demonstrated a marked gender difference and a dependence on the specific composition of the MLP diet [30, 31].

The most extensively studied postnatal outcome in MLP offspring is alterations in blood pressure, which on average is found to be elevated in offspring at an early age [16, 27, 32]. Elevations in blood pressure within MLP offspring are typically associated with reduced nephron numbers and reductions in kidney size [33, 34]. Another well documented outcome in MLP offspring are alterations in glucose homeostasis. Typically, fasting plasma insulin and glucose levels are reduced in young MLP offspring, suggesting an improvement in insulin sensitivity in early adulthood [35, 36]. However, these offspring demonstrate a swift decline in glucose homeostasis leading to glucose intolerance in later adulthood [35, 36]. In male offspring, impaired glucose tolerance develops into insulin resistant diabetes associated with hyperinsulinaemia by an age of 17 months [35]. While in female offspring, glucose intolerance develops much later, at 21 months, and is associated with relative hypoinsulinaemia [37]. Postnatal alterations in glucose homeostasis are typically associated with impaired pancreatic development [26] and age-dependent changes in peripheral insulin sensitivity [38, 39]. More recently, studies have shown that offspring exposed to maternal protein restriction during pregnancy exhibit a reduced life span [40] and sex-specific changes in appetite and food preference [41]. Like other restrictive nutritional approaches, the MLP model has experienced criticisms about its relevance in today's modern society. However, proponents of this approach highlight its relevance to cultures or socio-economic groups for whom dietary protein sources are scarce or expensive [31].

Maternal High Fat Feeding
In recognition of the criticisms levelled against restrictive nutritional studies, models of maternal overnutrition during pregnancy have also been developed. Currently maternal overnutrition is typically induced by maternal high fat feeding during pregnancy, diets used in such studies typically contain between 20 and 40% saturated animal fat [42–44]. In contrast to models of prenatal nutritional restriction, the offspring of high-fat fed mothers do not demonstrate FGR and are of normal or slightly higher birth weight [42, 43]. Notably, prenatal exposure to maternal high-fat feeding is associated with a number of adverse postnatal outcomes which share some similarities to those observed in restrictive animal studies. Adult offspring exposed to maternal high-fat feeding demonstrate higher blood glucose [43–45], hypertriglyceridaemia [43, 45], hyperinsulinaemia [45, 46], glucose intolerance [45], increased adiposity [43, 47], and gender-specific hypertension [42]. In contrast to the studies using nutritional restriction, studies of maternal overnutrition examine causal mechanisms that some investigators consider more relevant to the present dietary habits of western societies, specifically addressing the increasing availability of high-fat food and the epidemic rise of diet-induced obesity (DIO) [48].

Hormonal Interventions during Early Development

Various hormones play important roles in the regulation of early growth and development. As such changes in the levels of key hormones during pregnancy have been suggested to act as a central mechanism through which environmental stimuli alter fetal growth and development, thereby programming future disease risk. In the rat, prenatal overexposure to synthetic glucocorticoids has been developed as the most common hormonal DOHaD model [49]. Fetal glucocorticoid over-exposure is typically brought about by the treatment of pregnant rat dams with dexamethasone (Dex) [49], a synthetic glucocorticoid which freely crosses the placenta. In the rat, fetal glucocorticoid exposure causes FGR [49, 50] and is associated with a postnatal elevation in blood pressure [49], enhanced activation of the HPA axis [51], glucose intolerance [52], and fasting hyperinsulinaemia and hyperglycaemia [52]. One of the most notable features of glucocorticoid studies is the presence of a clearly defined window of susceptibility in the last trimester of pregnancy. Prenatal glucocorticoid exposure outside this specific window fails to have any effects on fetal growth and does not appear to have any adverse long-term postnatal consequences [51, 53–55]. More recently, prenatal exposure to inflammatory cytokines has been developed as another hormonal DOHaD model. In this model, pregnant rats are treated with either interleukin-6 (IL-6) or tumour necrosis factor-α (TNF-α) during pregnancy [56, 57]. In contrast to most other developmental models, prenatal cytokine exposure does not cause FGR [56, 57]. However, offspring exposed to inflammatory cytokines during early development develop obesity [56], gender-specific alterations in insulin sensitivity and secretion [56], hypertension [57, 58], impairments in spatial learning [59] and a dysregulation of the HPA axis [58]. This hormonal approach is believed to model some of the endocrine aspects of a maternal infection during pregnancy which is associated with an increased production of inflammatory cytokines like TNF-α [56]. Recent studies suggest that the placenta may be permeable to these inflammatory factors leaving the fetus vulnerable to an inappropriate exposure to inflammatory cytokines during development [60].

Functional Interventions during Pregnancy

The concept of the 'fetal supply line' encompasses the highly integrated supply system operating between the mother, placenta and fetus [62]. The nutritional studies described previously focus on an attempt to reduce nutritional inputs into this supply system. An alternative approach involves a direct impairment of the functional transport capacity of the fetal supply line. The central organ of the fetal supply line is the placenta, and in the clinical setting placental dysfunction has long been a well-established cause of fetal growth impairment [63]. Well before the inception of the developmental origins hypothesis, bilateral uterine artery ligation was developed in the rat to model the immediate neonatal outcomes of placental insufficiency [64]. More recent experimental studies investigating the developmental origins hypothesis have now adopted this experimental approach to study the long-term postnatal outcomes of placental insufficiency. Proponents of this approach stress its relevance to western societies where placental dysfunction is recognised as the most likely cause of fetal growth failure and low birth weight babies [10]. Studies in the rat using this experimental approach have clearly shown that placental insufficiency is responsible for fetal growth retardation, reduced birth weight and adverse postnatal outcomes [65, 66]. In adulthood, offspring exposed to placental insufficiency develop impairments in renal development [65], elevated blood pressure [66], impaired glucose tolerance [67], and demonstrate a progressive development of obesity and type II diabetes [67]. Critics of nutritional models have highlighted placental insufficiency as the most clinically relevant approach to the study of fetal growth impairment and the DOHaD hypothesis in most modern Western societies [10].

Biological Mechanisms

Experimental animal studies have clearly shown that adverse prenatal environmental stimuli can cause long-term physiological changes in the developing fetus that manifest in adulthood as significant disease risk factors. Current research now focuses on identifying the specific molecular mechanisms which are responsible for the connection between early-life experiences and future diseases processes. Experimental studies presently focus on the events that occur during the prenatal developmental period as this is the time of greatest biological plasticity and therefore greatest vulnerability to adverse stimuli.

Two general developmental mechanisms have been proposed to describe the manner in which adverse stimuli may affect prenatal development; these are developmental disruption and adaptation [68, 69]. Developmental disruption involves a direct teratogenic response to highly averse or toxic environmental stimuli, arresting or disrupting normal developmental processes. This type of disruptive response is believed to result in immediate physical malformations or functional deficits which are fundamentally maladaptive in nature [70]. In contrast, developmental adaptation involves a more subtle homeostatic alteration of development, reorganising developmental processes and physiological resources in a manner that maximises the chances of immediate survival in unfavourable environmental conditions [69]. While hypothetically providing a direct advantage necessary for immediate survival, such developmental adaptations may have maladaptive outcomes in the future when exposed to different postnatal environmental conditions during adult life [68].

Three molecular mechanisms have been proposed to explain how disruptive and adaptive developmental responses can permanently reshape the physiological form and function of a conceptus. These three mechanisms are: the remodelling of tissue development, the re-setting of homeostatic endocrine axes, and the permanent alteration of gene expression [71]. Biological examples of these three molecular mechanisms have been identified across the wide range of different rat models that have been developed to investigate the DOHaD hypothesis.

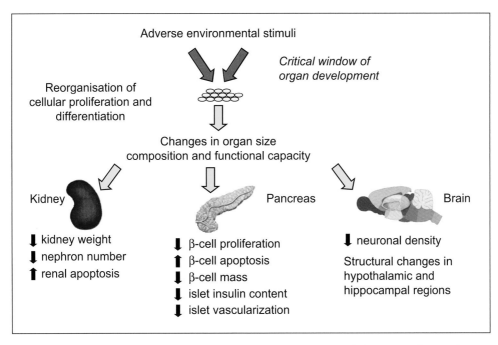

Fig. 2. Summary of tissue re-modelling mechanisms, summarising well documented examples.

Tissue Remodelling

Tissue remodelling involves developmental changes that alter the physical size and composition of tissues and organs through direct changes in cell number and type. The development of most major tissues is most sensitive to the effects of adverse stimuli during the embryonic and fetal periods of development. At this time, embryonic progenitor cells undergo co-ordinated waves of migration, proliferation and differentiation as part of the process of organogenesis. Exposure to adverse environmental stimuli during these sensitive periods can irreversibly alter tissue structure and function by disrupting or reorganising the tissue specific patterns of cellular proliferation and differentiation (fig. 2). It is possible for this response to arise from teratogenic stimuli [70] or as part of a homeostatic adaptation conserving and redistributing limited resources in a sub-optimal environmental setting [71].

Many examples of tissue remodelling have been identified among the different rat models described in this work. One of the most well-documented examples occurs in the kidney, where many studies have identified a permanent remodelling of kidney size and cellular composition. In offspring exposed to maternal protein restriction during pregnancy, kidney mass [33] and nephron number [33, 72] are both reduced while renal apoptosis is elevated [73]. Likewise, placental insufficiency by uterine artery ligation and maternal iron restriction during pregnancy have also been implicated in a permanent reduction of renal mass and nephron number [65, 74]

Another common example of tissue remodelling occurs in the pancreas where islet cell development appears to be particularly sensitive to adverse prenatal environmental stimuli. In MLP rat studies, offspring of protein-restricted mothers exhibit significant alterations in pancreas morphology including reduced β-cell mass, increased β-cell apoptosis, reduced β-cell proliferation, reduced islet insulin content and impaired islet vascularisation [26, 75]. Likewise, global maternal undernutrition has also been implicated in the induction of similar defects in pancreas development, reducing offspring's β-cell mass and islet insulin content [14, 18, 76].

The organisational development of neuronal nuclei within the brain also appears to be susceptible to adverse prenatal environmental influences. Significant alterations in neuronal density and the architecture of discrete hypothalamic nuclei have been observed in offspring exposed to maternal protein restriction [77] and gestational diabetes [78]. Equally, changes in hippocampal structure and morphology have also been documented in offspring overexposed to inflammatory cytokines [59] and excess glucocorticoids [53] during prenatal development.

Overall, the structural reorganisation of tissue development provides a common and well-defined mechanism by which adverse prenatal environmental stimuli can permanently alter tissue form and function in later life. Significant developmental changes in key organs like the kidney and pancreas have serious implications for future susceptibility to hypertension, diabetes and cardiovascular disease.

Endocrine Re-Setting

Endocrine factors play an important role in the regulation of most major physiological processes. Fetal growth and development is one such process organised by the complex interplay between the environment and the maternal and fetal endocrine signalling systems [79]. A re-setting or alteration of maternal/fetal endocrine function is therefore likely to be a key mechanism through which environmental stimuli act to influence prenatal growth and development [80]. Endocrine activity is tightly regulated by a complex network of environmental stimuli and feedback inputs which work to modulate hormone production, release, bio-activity and target tissue sensitivity. Thus, there are a number of regulatory levels at which adverse prenatal stimuli could act to permanently alter the regulation of maternal/fetal endocrine activity.

A re-setting of endocrine sensitivity has been implicated in a number of different endocrine systems including the renin-angiotensin system (RAS) [81], the HPA axis [21], and the insulin endocrine axis [67]. The RAS mediates the regulation of blood pressure, fluid homeostasis and plays a central role in kidney development. In offspring exposed to maternal protein restriction RAS activity is suppressed during fetal development [27] but enhanced postnatally in adulthood [81]. The suppression of fetal RAS activity is implicated in a failure of nephrogenesis during prenatal development

[27], while the postnatal upregulation of RAS provides a mechanism for the development of hypertension in adult offspring [81].

The HPA axis is involved in the regulation of a diverse number of biological processes but has a key role in mediating stress responses. In this context, it is not surprising that a number of different experimental rat studies have identified a re-setting of the sensitivity of the HPA axis in offspring exposed to a variety of adverse prenatal environmental stimuli. Studies using the MLP diet [82], 50% maternal undernutrition [21], maternal glucocorticoid exposure [83] and maternal cytokine exposure [58] have all shown effects on the HPA axis which appear to indicate a reduction in feedback inhibition at the level of the hypothalamus and conversely an enhancement of HPA axis activity. The hypothalamus is a central regulatory hub for a number of different endocrine systems, any possible alterations in the sensitivity of this hub will have many follow-on effects on a number of different hypothalamic endocrine axes [84].

The insulin endocrine axis plays a central role in the regulation of fetal growth during prenatal development and glucose homeostasis in adulthood. Across various rat studies, alterations in pancreatic function and peripheral insulin sensitivity emerge as a common pathos, indicating a unique sensitivity of the insulin endocrine axis to adverse prenatal stimuli [85]. Studies in the MLP model have found that young offspring demonstrate improved glucose tolerance and insulin sensitivity. However, these offspring experience a rapid age-dependent loss of glucose tolerance which is associated with β-cell exhaustion in females, and increasing insulin resistance in males [36]. A similar phenotypic progression is observed in studies using uterine artery ligation to induce placental insufficiency, with both male and female offspring experiencing an age-dependent decline in glucose tolerance and insulin sensitivity, accompanied by a loss of pancreatic β-cell mass [67]. Conversely, a recent study using maternal undernutrition during pregnancy has shown evidence, in adult offspring, of chronic insulin hypersecretion, which appears to be, associated with an enhancement of hepatic insulin signalling and increased hepatic de novo lipogenesis [23].

Notably, most of the examples of endocrine re-setting are associated with alterations in the activity of endocrine organs and target tissues. In this context, the remodelling of endocrine organs provides a plausible mechanism involved in the re-setting of endocrine activity. Conversely, endocrine signalling has a well established role in regulating tissue development and growth, thus alterations in endocrine activity can, in turn play a role in the remodelling of tissue development. In this context, tissue remodelling and endocrine re-setting can be seen as interdependent mechanisms acting in a concerted manner to shape the development and future physiology of the conceptus.

Changes in Gene Expression

An individual's genetic inheritance provides the basic plans for its future growth and development. However, the extent to which these plans are carried out is determined

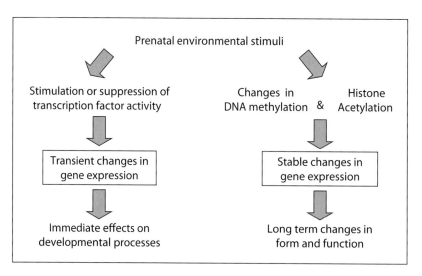

Fig. 3. Diagram detailing two possible mechanisms which may act to alter gene expression in offspring exposed to adverse prenatal environmental stimuli.

by the level of constraint imposed by the uterine environment [62]. Therefore, changes in gene expression represent one of the most fundamental mechanisms by which prenatal environmental stimuli can exert a direct effect on early growth and development [80]. Prenatal environmental influences may effect gene expression in at least two distinct ways (fig. 3). Firstly, gene expression may be transiently altered by alterations in the activity of transcription factors, gene promoters, and gene repressors. Secondly, environmental stimuli may induce more stable, heritable alterations in gene expression through the epigenetic modification of DNA and associated histone proteins. In the context of the DOHaD hypothesis epigenetic changes in gene expression represent a central molecular mechanism through which environmental influence may permanently alter the future physiological function of the conceptus. With regard to the biological mechanisms discussed previously, changes in gene expression may be both a cause and a consequence of tissue remodelling and endocrine re-setting.

Epigenetic changes in gene expression encompass the processes of DNA methylation and histone acetylation. DNA methylation involves the covalent addition of methyl groups to cytosine DNA residues [86]. Typically, promoter hypermethylation is associated with gene silencing, while hypomethylation is associated with active gene expression [87]. Likewise, histone acetylation involves the addition of an acetyl group to lysine histone residues [88]. Changes in histone acetylation affect chromatin conformation which in turn also affects gene expression. Typically, histone acetylation is associated with transcriptionally active gene loci, characterised by an extended chromatin conformation [88]. The epigenetic patterning of gene expression is established during gametogenesis and early embryogenesis and is particularly sensitive to disruptive environmental influences during this time [89].

Many studies have found changes in gene expression in both fetal and adult tissues of offspring exposed to adverse prenatal environmental stimuli [52, 72, 90–92]. However, it is only recently that investigators have begun to explore the epigenetic basis of these changes. Recent studies using the uterine artery ligation model have found that uteroplacental insufficiency can permanently alter DNA methylation patterns [93], histone acetylation [94] and chromatin structure [95], in the tissues of adult offspring. In the kidney, uteroplacental insufficiency is associated with the specific hypomethylation of the p53 gene, which is subsequently associated with an increase in p53 mRNA expression and an enhancement of renal apoptosis [93]. In the fetal brain, uteroplacental insufficiency is also associated with global hypomethylation and a reduction in histone acetylation [95]. Changes in the DNA methylation patterns have also been identified in the offspring of protein restricted mothers. One study demonstrates hepatic hypomethylation, associated with an increase in peroxisome proliferators-activated receptor-α (PPAR-α) and glucocorticoid receptor (GR) gene expression in the liver of MLP offspring [96].

Distinct Prenatal Pathways to Obesity

Over recent years, a number of interrelated environmental factors have been identified that lead to obesity development. First, as discussed it is now widely accepted that environmental changes during foetal development play a key role in determining susceptibility to obesity and metabolic disease in adult life [71]. Secondly, there is increasing evidence that a range of postnatal environmental conditions amplify the risk of metabolic disorders in adult life [97]. Such factors in rodent studies include neonatal overnutrition through either reduced litter size or overnutrition from lactating obese dams [97] and post-weaning diet-induced obesity [98].

It has recently been identified that while both diet-induced obesity (via a postnatal high-fat diet) and prenatally induced obesity (via maternal undernutrition during pregnancy) may appear phenotypically similar they have clearly different underlying mechanisms and metabolic outcomes [23]. Insulin secretion and insulin action clearly differ between these two obese groups. In adult life, FGR offspring showed elevated plasma fasting insulin levels with parallel increases in C-peptide secretion [23] providing evidence that increases in plasma insulin observed are the result of insulin hypersecretion from the pancreas and not an impairment of insulin uptake in peripheral tissues such as liver and muscle as observed in the insulin resistant state. There was no evidence of a reduction in insulin sensitivity in the FGR offspring compared to control animals on the same nutritional regime when assessed by hyperinsulinaemic-euglycaemic clamp [23]. In contrast, DIO rats show no elevations in insulin secretion but clear whole body reduction in insulin sensitivity [23]. Transport of glucose across the cell membrane is considered to be the rate-controlling step of insulin-regulated glucose metabolism in muscle and liver [99]. Insulin regulates glucose transport in muscle and hepatic tissue through activation of insulin receptor substrate-dependent phosphatidylinositol (PI)-3 kinase. Downstream effectors of PI3 kinases are postulated to mediate changes in levels of atypical protein kinase C ζ (PKC ζ) in the regulation of insulin-induced increases in glucose transport [100]. Increased hepatic PKC ζ protein concentration is evident in FGR offspring [23]. A recent report which shows that PKC ζ can increase insulin internalisation in vitro and thus insulin action within the liver [101]. Furthermore, linked increases in hepatic and muscle glycogen stores are evident in FGR offspring only [23]. In contrast, DIO and whole body insulin resistance was associated with a downregulation of hepatic PKC ζ [23] and type 2 diabetes sufferers exhibit reduced muscular glycogen storage [108]. It has been suggested that downregulation of the PKC ζ pathway plays an important role in the pathogenesis of insulin resistance in type 2 diabetes [102].

A link between hepatic and muscular triglyceride storage and insulin resistance has been widely accepted. Further, differences in fat storage have been identified between FGR obese and DIO rats. In FGR rats, obesity is associated with an increase in fat synthesis through elevations in hepatic fatty acid synthase levels and storage of this energy in peripheral adipose tissue sites not organs [23]. In contrast, DIO is associated with a reduction in hepatic fat synthesis. The observation, in the insulin resistant DIO rats of pathophysiological lipid deposition in non-adipose tissues, is in agreement with lipotoxic theory of obesity development [103].

Previous studies investigating the effects of FGR on insulin sensitivity in adult life have used sporadic plasma insulin measurements or glucose tolerance tests as a proxy of insulin sensitivity. From these studies, it was generally accepted that the hyperinsulinism observed in FGR offspring was evidence of underlying insulin resistance [104]. However, the notion that elevated circulating insulin or abnormal glucose tolerance tests always represent a reduction in insulin sensitivity is challenged by this recent animal study, and now supported by a growing number of clinical investigations [105]. Of particular concern, small-for-gestational-age children who have experienced intrauterine growth restriction are often classified as insulin resistant when assessed by the HOMA method [106]. Interestingly, the mechanisms responsible for 'catch-up growth' during the early postnatal period are thought to influence associations between FGR and risks for subsequent obesity development [107]. Our experimental approach clearly delineates that FGR offspring (in this case via maternal undernutrition) show insulin hypersecretion with maintained insulin sensitivity, while insulin resistance is induced by hypercaloric nutrition after birth. Therefore, it is feasible that an increase in insulin secretion and enhanced insulin action in FGR offspring, which would support catch-up growth, could be masked, under obesogenic conditions when insulin resistance develops with a high fat diet. The nature of this paradox, based on postnatal hypercaloric HF-nutrition, may be a reason, at least in part, why a number of studies of FGR offspring, in both clinical and animal settings, have been classified as insulin resistant.

Summary and Conclusions

A large number of epidemiological studies have described a significant relationship between markers of fetal growth restriction and an increased susceptibility to the development of metabolic disease in later adult life. The DOHaD hypothesis suggests that this relationship is a consequence of the developmental adaptations a fetus makes in response to adverse prenatal environmental stimuli. In support of this hypothesis, experimental animal studies have clearly shown that exposure to adverse prenatal stimuli can induce permanent physiology changes which form significant risk factors for the development of metabolic and cardiovascular abnormalities in adult life. Many different experimental approaches have been used to investigate the DOHaD hypothesis, most have attempted to model common events that adversely influence human pregnancies such as undernutrition, placental dysfunction, maternal diabetes/obesity and maternal infection. Across the wide range of different models, fetal growth restriction, characterised by a reduced birth weight, appears to be closely associated with common disturbances in kidney function and metabolic homeostasis. Altogether, there is significant evidence to suggest that an increased susceptibility to cardiovascular and metabolic disease is a common consequence of fetal growth restriction.

References

1 Barker DJP: Mothers, Babies and Disease in Later Life. London, BMJ Publishing Group, 1994.
2 Barker DJ, et al: Growth in utero, blood pressure in childhood and adult life, and mortality from cardiovascular disease. BMJ 1989;298:564–567.
3 Barker DJ, et al: Weight in infancy and death from ischaemic heart disease. Lancet 1989;ii:577–580.
4 Barker DJ, et al: Fetal and placental size and risk of hypertension in adult life. BMJ 1990;301:259–262.
5 Hales CN, et al: Fetal and infant growth and impaired glucose tolerance at age 64. BMJ 1991;303:1019–1022.
6 Barker DJ, et al: Type 2 (non-insulin-dependent) diabetes mellitus, hypertension and hyperlipidaemia (syndrome X): relation to reduced fetal growth. Diabetologia 1993;36:62–67.
7 Lucas A, Fewtrell MS, Cole TJ: Fetal origins of adult disease-the hypothesis revisited. BMJ 1999;319: 245–249.
8 Joseph KS, Kramer MS: Review of the evidence on fetal and early childhood antecedents of adult chronic disease. Epidemiol Rev 1996;18:158–174.

9 Hattersley AT, Tooke JE: The fetal insulin hypothesis: an alternative explanation of the association of low birthweight with diabetes and vascular disease. Lancet 1999;353:1789–1792.
10 Henriksen T, Clausen T: The fetal origins hypothesis: placental insufficiency and inheritance versus maternal malnutrition in well-nourished populations. Acta Obstet Gynecol Scand 2002;81:112–114.
11 Huxley R, Neil A, Collins R: Unravelling the fetal origins hypothesis: is there really an inverse association between birthweight and subsequent blood pressure? Lancet 2002;360:659–665.
12 Schluchter MD: Publication bias and heterogeneity in the relationship between systolic blood pressure, birth weight, and catch-up growth: a meta analysis. J Hypertens 2003;21:273–279.
13 Woodall SM, et al: Chronic maternal undernutrition in the rat leads to delayed postnatal growth and elevated blood pressure of offspring. Pediatr Res 1996;40:438–443.

14 Garofano A, Czernichow P, Breant B: Beta-cell mass and proliferation following late fetal and early post-natal malnutrition in the rat. Diabetologia 1998;41: 1114–1120.

15 van Marthens E: Alterations in the rate of fetal and placental development as a consequence of early maternal protein/calorie restriction. Biol Neonate 1977;31:324–332.

16 Langley-Evans SC, Phillips GJ, Jackson AA: In utero exposure to maternal low protein diets induces hypertension in weaning rats, independently of maternal blood pressure changes. Clin Nutr 1994;13: 319–324.

17 Ozaki T, et al: Dietary restriction in pregnant rats causes gender-related hypertension and vascular dysfunction in offspring. J Physiol (Lond) 2001;530: 141–152.

18 Garofano A, Czernichow P, Breant B: Postnatal somatic growth and insulin contents in moderate or severe intrauterine growth retardation in the rat. Biol Neonate 1998;73:89–98.

19 Garofano A, Czernichow P, Breant B: Effect of age-ing on beta-cell mass and function in rats malnour-ished during the perinatal period. Diabetologia 1999;42:711–718.

20 Chen CM, Wang LF, Su B: Effects of maternal undernutrition during late gestation on the lung surfactant system and morphometry in rats. Pediatr Res 2004;56:329–335.

21 Lesage J, et al: Maternal undernutrition during late gestation induces fetal overexposure to glucocorti-coids and intrauterine growth retardation, and dis-turbs the hypothalamo-pituitary adrenal axis in the newborn rat. Endocrinology 2001;142:1692–1702.

22 Vickers MH, et al: Fetal origins of hyperphagia, obe-sity, and hypertension and postnatal amplification by hypercaloric nutrition. Am J Physiol Endocrinol Metab 2000;279:E83–E87.

23 Thompson NM, et al: Prenatal and postnatal path-ways to obesity: different underlying mechanisms, different metabolic outcomes. Endocrinology 2007; 148:2345–2354.

24 Yajnik CS: Early life origins of insulin resistance and type 2 diabetes in India and other Asian countries. J Nutr 2004;134:205–210.

25 Holness MJ, Sugden MC: Antecedent protein rest-riction exacerbates development of impaired insulin action after high-fat feeding. Am J Physiol 1999; 276:E85–E93.

26 Snoeck A, et al: Effect of a low protein diet during pregnancy on the fetal rat endocrine pancreas. Biol Neonate 1990;57:107–118.

27 Woods LL, et al: Maternal protein restriction sup-presses the newborn renin-angiotensin system and programs adult hypertension in rats. Pediatr Res 2001;49:460–467.

28 Desai M, et al: Organ-selective growth in the off-spring of protein-restricted mothers. Br J Nutr 1996;76:591–603.

29 Langley SC, Browne RF, Jackson AA: Altered glu-cose tolerance in rats exposed to maternal low pro-tein diets in utero. Comp Biochem Physiol Physiol 1994;109:223–229.

30 Langley-Evans SC: Critical differences between two low protein diet protocols in the programming of hypertension in the rat. Int J Food Sci Nutr 2000;51: 11–17.

31 Langley-Evans SC: Developmental programming of health and disease. Proc Nutr Soc 2006;65:97–105.

32 Manning J, Vehaskari VM: Low birth weight-associ-ated adult hypertension in the rat. Pediatr Nephrol 2001;16:417–422.

33 Langley-Evans SC, Welham SJ, Jackson AA: Fetal exposure to a maternal low protein diet impairs nephrogenesis and promotes hypertension in the rat. Life Sci 1999;64:965–974.

34 Jones SE, et al: Intra-uterine environment influ-ences glomerular number and the acute renal adap-tation to experimental diabetes. Diabetologia 2001; 44:721–728.

35 Petry CJ, et al: Diabetes in old male offspring of rat dams fed a reduced protein diet. Int J Exp Diabetes Res 2001;2:139–143.

36 Hales CN, et al: Fishing in the stream of diabetes: from measuring insulin to the control of fetal orga-nogenesis. Biochem Soc Trans 1996;24:341–350.

37 Petry CJ, et al: Effects of early protein restriction and adult obesity on rat pancreatic hormone con-tent and glucose tolerance. Horm Metab Res 2000; 32:233–239.

38 Ozanne SE, et al: Altered muscle insulin sensitivity in the male offspring of protein-malnourished rats. Am J Physiol 1996;271(6 Pt 1):E1128–E1134.

39 Ozanne SE, et al: Altered regulation of hepatic glu-cose output in the male offspring of protein-mal-nourished rat dams. Am J Physiol 1996;270(4 Pt 1): E559–E564.

40 Aihie Sayer A, et al: Prenatal exposure to a maternal low protein diet shortens life span in rats. Gerontol-ogy 2001;47:9–14.

41 Bellinger L, Lilley C, Langley-Evans SC: Prenatal expo-sure to a maternal low-protein diet programmes a preference for high-fat foods in the young adult rat. Br J Nutr 2004;92:513–520.

42 Khan IY, et al: Gender-linked hypertension in off-spring of lard-fed pregnant rats. Hypertension 2003; 41:168–175.

43 Guo F, Jen KL: High-fat feeding during pregnancy and lactation affects offspring metabolism in rats. Physiol Behav 1995;57:681–686.

44 Cerf ME, et al: Hyperglycaemia and reduced glu-cokinase expression in weanling offspring from dams maintained on a high-fat diet. Br J Nutr 2006; 95:391–396.

45 Srinivasan M, et al: Maternal high-fat diet con-sumption results in fetal malprogramming predis-posing to the onset of metabolic syndrome-like phenotype in adulthood. Am J Physiol Endocrinol Metab 2006;291:E792–E799.

46 Khan IY, et al: A high-fat diet during rat pregnancy or suckling induces cardiovascular dysfunction in adult offspring. Am J Physiol Regul Integr Comp Physiol 2005;288:R127–R133.

47 Buckley AJ, et al: Altered body composition and metabolism in the male offspring of high fat-fed rats. Metabolism 2005;54:500–507.

48 Levin BE: Metabolic imprinting: critical impact of the perinatal environment on the regulation of energy homeostasis. Philos Trans R Soc Lond [B] 2006;361:1107–1121.

49 Benediktsson R, et al: Glucocorticoid exposure in utero: new model for adult hypertension. Lancet 1993;341:339–341.

50 Lindsay RS, et al: Prenatal glucocorticoid exposure leads to offspring hyperglycaemia in the rat: studies with the 11-beta-hydroxysteroid dehydrogenase inhibitor carbenoxolone. Diabetologia 1996;39: 1299–1305.

51 Welberg LA, Seckl JR, Holmes MC: Inhibition of 11beta-hydroxysteroid dehydrogenase, the foeto-placental barrier to maternal glucocorticoids, per-manently programs amygdala GR mRNA expression and anxiety-like behaviour in the off-spring. Eur J Neurosci 2000;12:1047–1054.

52 Nyirenda MJ, et al: Glucocorticoid exposure in late gestation permanently programs rat hepatic phos-phoenolpyruvate carboxykinase and glucocorticoid receptor expression and causes glucose intolerance in adult offspring. J Clin Invest 1998;101:2174–2181.

53 Levitt NS, et al: Dexamethasone in the last week of pregnancy attenuates hippocampal glucocorticoid receptor gene expression and elevates blood pres-sure in the adult offspring in the rat. Neuroendo-crinology 1996;64:412–418.

54 Seckl JR, Cleasby M, Nyirenda MJ: Glucocorticoids, 11-beta-hydroxysteroid dehydrogenase, and fetal programming. Kidney Int 2000;57:1412–1417.

55 Woods LL, Weeks DA: Prenatal programming of adult blood pressure: role of maternal corticos-teroids. Am J Physiol Regul Integr Comp Physiol 2005;289:R955–R962.

56 Dahlgren J, et al: Prenatal cytokine exposure results in obesity and gender-specific programming. Am J Physiol Endocrinol Metab 2001;281:E326–E334.

57 Samuelsson AM, et al: Prenatal exposure to inter-leukin-6 results in hypertension and alterations in the renin-angiotensin system of the rat. J Physiol 2006;575(Pt 3):855–867.

58 Samuelsson AM, et al: Prenatal exposure to inter-leukin-6 results in hypertension and increased hypothalamic-pituitary-adrenal axis activity in adult rats. Endocrinology 2004;145:4897–4911.

59 Samuelsson AM, et al: Prenatal exposure to inter-leukin-6 results in inflammatory neurodegeneration in hippocampus with NMDA/GABA(A) dysregula-tion and impaired spatial learning. Am J Physiol Regul Integr Comp Physiol 2006;290:R1345–R1356.

60 Dahlgren J, et al: Interleukin-6 in the maternal cir-culation reaches the rat fetus in mid-gestation. Pediatr Res 2006;60:147–151.

61 Boggess KA: Pathophysiology of preterm birth: emerging concepts of maternal infection. Clin Perinatol 2005;32:561–569.

62 Harding JE: Nutrition and growth before birth. Asia Pacif J Clin Nutr 2003;12(suppl):S28.

63 Baschat AA, Hecher K: Fetal growth restriction due to placental disease. Semin Perinatol 2004;28:67–80.

64 Wigglesworth JS: Fetal growth retardation. Animal model: uterine vessel ligation in the pregnant rat. Am J Pathol 1974;77:347–350.

65 Merlet-Benichou C, et al: Intrauterine growth retar-dation leads to a permanent nephron deficit in the rat. Pediatr Nephrol 1994;8:175–180.

66 Jansson T, Lambert GW: Effect of intrauterine growth restriction on blood pressure, glucose toler-ance and sympathetic nervous system activity in the rat at 3–4 months of age. J Hypertens 1999;17: 1239–1248.

67 Simmons RA, Templeton LJ, Gertz SJ: Intrauterine growth retardation leads to the development of type 2 diabetes in the rat. Diabetes 2001;50:2279–2286.

68 Gluckman PD, Hanson MA, Pinal C: The develop-mental origins of adult disease. Maternal Child Nutr 2005;1:130–141.

69 Bateson P, et al: Developmental plasticity and human health. Nature 2004;430:419–421.

70 Plagemann A: Perinatal programming and func-tional teratogenesis: impact on body weight regula-tion and obesity. Physiol Behav 2005;86:661–668.

71 Gluckman PD, Hanson MA: Living with the past: evolution, development, and patterns of disease. Science 2004;305:1733–1736.

72 McMullen S, Gardner DS, Langley-Evans SC: Prenatal programming of angiotensin II type 2 receptor expression in the rat. Br J Nutr 2004;91: 133–140.

73 Welham SJ, Wade A, Woolf AS: Protein restriction in pregnancy is associated with increased apoptosis of mesenchymal cells at the start of rat metanephrogenesis. Kidney Int 2002;61:1231–1242.

74 Lisle SJ, et al: Effect of maternal iron restriction during pregnancy on renal morphology in the adult rat offspring. Br J Nutr 2003;90:33–39.

75 Dahri S, et al: Islet function in offspring of mothers on low-protein diet during gestation. Diabetes 1991;40(suppl 2):115–120.

76 Garofano A, Czernichow P, Breant B: In utero undernutrition impairs rat beta-cell development. Diabetologia 1997;40:1231–1234.

77 Plagemann A, et al: Hypothalamic nuclei are malformed in weanling offspring of low protein malnourished rat dams. J Nutr 2000;130:2582–2589.

78 Plagemann A, et al: Malformations of hypothalamic nuclei in hyperinsulinemic offspring of rats with gestational diabetes. Dev Neurosci 1999;21:58–67.

79 Fowden AL: Endocrine regulation of fetal growth. Reprod Fertil Dev 1995;7:351–363.

80 Gluckman P, Hanson M: The Fetal Matrix: Evolution, Development and Disease. Cambridge, Cambridge University Press, 2005.

81 Langley-Evans SC, et al: Intrauterine programming of hypertension: the role of the renin-angiotensin system. Biochem Soc Trans 1999;27:88–93.

82 Langley-Evans SC, Gardner DS, Jackson AA: Maternal protein restriction influences the programming of the rat hypothalamic-pituitary-adrenal axis. J Nutr 1996;126:1578–1585.

83 Shoener JA, Baig R, Page KC: Prenatal exposure to dexamethasone alters hippocampal drive on hypothalamic-pituitary-adrenal axis activity in adult male rats. Am J Physiol Regul Integr Comp Physiol 2006;290:R1366–R1373.

84 Slone-Wilcoxon J, Redei EE: Maternal-fetal glucocorticoid milieu programs hypothalamic-pituitary-thyroid function of adult offspring. Endocrinology 2004;145:4068–4072.

85 Fernandez-Twinn DS, Ozanne SE: Mechanisms by which poor early growth programs type-2 diabetes, obesity and the metabolic syndrome. Physiol Behav 2006;88:234–243.

86 Singal R, Ginder GD: DNA methylation. Blood 1999;93:4059–4070.

87 Newell-Price J, Clark AJ, King P: DNA methylation and silencing of gene expression. Trends Endocrinol Metab 2000;11:142–148.

88 Peterson CL, Laniel MA: Histones and histone modifications. Curr Biol 2004;14:R546–R551.

89 Jaenisch R, Bird A: Epigenetic regulation of gene expression: how the genome integrates intrinsic and environmental signals. Nat Genet 2003;33(suppl): 245–254.

90 Bertram C, et al: The maternal diet during pregnancy programs altered expression of the glucocorticoid receptor and type 2 11beta-hydroxysteroid dehydrogenase: potential molecular mechanisms underlying the programming of hypertension in utero. Endocrinology 2001;142:2841–2853.

91 Guan H, et al: Adipose tissue gene expression profiling reveals distinct molecular pathways that define visceral adiposity in offspring of maternal protein-restricted rats. Am J Physiol Endocrinol Metab 2005; 288:E663–E673.

92 Lopes Da Costa C, Sampaio De Freitas M, Sanchez Moura A: Insulin secretion and GLUT-2 expression in undernourished neonate rats. J Nutr Biochem 2004;15:236–241.

93 Pham TD, et al: Uteroplacental insufficiency increases apoptosis and alters p53 gene methylation in the full-term IUGR rat kidney. Am J Physiol Regul Integr Comp Physiol 2003;285:R962–R970.

94 MacLennan NK, et al: Uteroplacental insufficiency alters DNA methylation, one-carbon metabolism, and histone acetylation in IUGR rats. Physiol Genomics 2004;18:43–50.

95 Ke X, et al: Uteroplacental insufficiency affects epigenetic determinants of chromatin structure in brains of neonatal and juvenile IUGR rats. Physiol Genomics 2006;25:16–28.

96 Lillycrop KA, et al: Dietary protein restriction of pregnant rats induces and folic acid supplementation prevents epigenetic modification of hepatic gene expression in the offspring. J Nutr 2005;135: 1382–1386.

97 Gorski JN, et al: Postnatal environment overrides genetic and prenatal factors influencing offspring obesity and insulin resistance. Am J Physiol Regul Integr Comp Physiol 2006;291:R768–R778.

98 Levin BE, Dunn-Meynell AA: Reduced central leptin sensitivity in rats with diet-induced obesity. Am J Physiol Regul Integr Comp Physiol 2002;283: R941–R948.

99 Ozanne SE, et al: Early growth restriction leads to down regulation of protein kinase C zeta and insulin resistance in skeletal muscle. J Endocrinol 2003;177: 235–241.

100 Sajan MP, Rivas J, Li P, Standaert ML, Farese RV: Repletion of atypical protein kinase C following RNA interference-mediated depletion restores insulin-stimulated glucose transport. J Biol Chem 2006;281:17466–17473.

101 Fiory F, Oriente F, Miele C, Romano C, Trencia A, Alberobello AT, Esposito I, Valentino R, Beguinot F, Formisano P: Protein kinase C-zeta and protein kinase B regulate distinct steps of insulin endocytosis and intracellular sorting. J Biol Chem 2004;279: 11137–11145.

102 Kim Y-B, et al: Insulin-stimulated protein kinase C {lambda}/{zeta} activity is reduced in skeletal muscle of humans with obesity and type 2 diabetes: reversal with weight reduction. Diabetes 2003;52:1935–1942.

103 Unger RH: The physiology of cellular liporegulation. Annu Rev Physiol 2003;65:333–347.

104 Benyshek DC, Johnston CS, Martin JF: Post-natal diet determines insulin resistance in fetally malnourished, low birthweight rats (F1) but diet does not modify the insulin resistance of their offspring (F2). Life Sci 2004;74:3033–3041.

105 Huang BW, et al: The effect of high-fat and high-fructose diets on glucose tolerance and plasma lipid and leptin levels in rats. Diabetes Obes Metab 2004;6:120–126.

106 Kind KL, et al: Effect of maternal feed restriction during pregnancy on glucose tolerance in the adult guinea pig. Am J Physiol Regul Integr Comp Physiol 2003;284:R140–R152.

107 Bazaes RA, et al: Determinants of insulin sensitivity and secretion in very-low-birth-weight children. J Clin Endocrinol Metab 2004;89:1267–1272.

108 Clines GW, et al: Impaired glucose transport as a cause of decreased insulin stimulated muscle glycogen synthesis in type 2 diabetes. N Engl J Med 1999; 341:240–246.

Associate Professor Bernhard H. Breier
Liggins Institute
University of Auckland, Private Bag 92019
Auckland 1023 (New Zealand)
Tel. +64 9 373 7599 (ext. 86442), Fax +64 9 373 7497, E-Mail bh.breier@auckland.ac.nz

Kiess W, Chernausek SD, Hokken-Koelega ACS (eds): Small for Gestational Age. Causes and Consequences.
Pediatr Adolesc Med. Basel, Karger, 2009, vol 13, pp 44–59

Molecular Genetic Disorders of Fetal Growth

Steven D. Chernausek

Children's Medical Research Institute (CMRI), Diabetes and Metabolic Research Program,
Department of Pediatrics, University of Oklahoma Health Sciences Center,
Oklahoma City, Okla., USA

Abstract

Our understanding of the control of fetal growth at the molecular level has increased greatly over recent decades. Genetic manipulation of mice and identification of specific gene defects in humans that result in intra-uterine growth retardation indicate major involvement of insulin and the insulin-like growth factor (IGF) axes. Defects in the genes that are necessary for insulin production or action as well as those of the IGF pathway have all been reported in humans and help define the specific roles these factors play in fetal growth and development. This chapter reviews the pathophysiology, clinical features, diagnostic measures, and management of patients with specific genetic causes of fetal growth retardation.

I often say to students: 'Most things that go wrong with your car make it run slower, not faster'. Similarly, most conditions that chronically perturb various physiologic functions also slow processes of growth. Since the diagnostic possibilities are vast, the clinician frequently must 'play detective' when faced with an abnormal growth pattern, either to come to a diagnosis or explain the cause of growth retardation. When fetal growth is impaired, arriving at a diagnosis is especially challenging. Though abnormal growth may be part of an identifiable syndrome, the exact mechanism restricting growth is rarely evident; often no specific diagnosis is reached.

The number of disorders for which the specific genetic basis is known has increased geometrically over the last decade. Many of these occur in the context of disordered fetal growth and are now diagnosable using molecular genetic techniques. Rather than compile all molecular genetic disorders that affect fetal growth (which would be incomplete by the time this comes to print), I will instead review the principle hormonal pathways known to control fetal growth and describe defined genetic disorders that impinge upon these pathways. In this way, the reader will gain an understanding of the processes that lead to abnormalities of fetal growth and have a foundation for the evaluation of affected

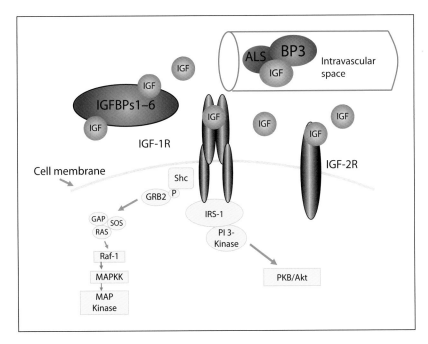

Fig. 1. Schematic of IGF action. IGF (either IGF-I or -II) circulates in a ternary complex with the acid labile subunit (ALS) and IGF-binding protein 3 (BP-3). The IGFBPs all bind IGFs and modulate IGF action at the tissue level. IGF signals via the type 1 IGF receptor (IGF-1R) via 2 main pathways involving AKT and MAP kinase. Not all intracellular elements are shown.

children now and in the future. For information concerning genetic disorders not covered here or defined subsequent to this writing, the reader is referred to the website Online Mendelian Inheritance in Man (OMIM, http://www.ncbi.nlm.nih.gov/sites/entrez?db = OMIM), an excellent and updated compendium of human genetic disorders.

The Insulin-Like Growth Factor Axis in Fetal Growth Control

The insulin-like growth factor (IGF) axis is the main controller of prenatal and postnatal growth and a major growth regulatory pathway within which genetic correlates of human disease have been identified. In mice, 85% of body size can be explained by the combined actions of growth hormone and the IGFs [1]. In humans, disorders that reduce IGF-I expression have profound effects on growth. What follows is a description of the components of the IGF growth control pathway, a review of the roles each play controlling general somatic fetal growth, and a discussion of a limited number of specific human genetic disorders.

The principal components of the IGF system are shown in figure 1. These consist of two ligands, IGF-I and IGF-II, six binding proteins (IGF-binding proteins 1–6), and

Table 1. GH/IGF pathway null mutant phenotypes

Gene knock-out	Birth weight (% normal)	Postnatal growth	Other features	References
GhR	100	very slow	very low circulating IGF-I	[1, 74]
Igf-I	60	very slow	decreased brain growth, normal placenta	[2, 5, 6]
Igf-II	60	normal	small placenta; imprinted gene	[7, 75]
Igf-I & II	30	NA	neonatal death	[6, 76]
Igf-1r	45	NA	neonatal death	[2]
Igf-2/M6p r	135	NA	perinatal death; imprinted gene	[3]
Irs-1	60	normal	insulin resistant	[77, 78]
Irs-2	90	90% of normal	hyperglycemia in neonates; overt DM by 10 weeks	[79]
Akt 1	80–90	70–80% normal	glucose tolerance normal	[10, 11]
Akt 2	90–100	mild deficit	hyperglycemia progressing to DM in males	[12]
Akt 3	100	normal	reduced brain size	[13]
Akt 1 & 2	53	NA	neonatal death; phenocopy of Igf-1r null mutants	[15]
Akt 2 & 3	100	75–85% of normal	males more growth retarded than females; glucose intolerant	[80]

two cell-surface receptors, the type I IGF receptor (IGF-1R) the IGF-II/mannose-6 phosphate receptor (IGF-2R). The IGF-1R is a disulphide-linked heterotetramer structurally related to the insulin receptor and mediates growth signals of IGFs. It contains a tyrosine-kinase domain that, when occupied by ligand, initiates a cascade of intracellular events shown diagrammatically in figure 1. These result in the phosphorylation of docking proteins and activation of additional intracellular kinases, which eventually activate genes that stimulate growth. In addition, the insulin receptor binds IGF-I at low affinity and can form hybrid receptors with the IGF-1R gene product.

The intracellular, postreceptor elements form a network involved in a variety of signaling systems. The relative and respective contributions of these components of fetal growth control have been elucidated through specific experiments wherein each has been eliminated by targeted gene deletion (these results are summarized in table 1). These studies reveal the following:

(1) The IGF-1R is the primary mediator of the growth-promoting effects of the IGFs. Mice rendered null for the IGF-1R are extremely small and usually die shortly after birth [2].

(2) The IGF-2R clears IGF-I and -II from the pericellular environment, thereby reducing IGF signaling. Thus, when the IGF-2R is knocked out, the result is an animal larger than normal, presumably due to the increased abundance of IGF now interacting with the type 1 IGF receptor [3]. Though there is some evidence for signaling via the IGF-2R [4], it is not yet clear how this applies to control of fetal growth.

(3) IGF-I is the dominant factor in growth regulation, mediating growth-promoting actions of growth hormone (GH). IGF-I-null mice are born small and show severe postnatal growth retardation [5, 6]. IGF-II is also important, but appears be more involved in fetal growth. Deletion of IGF-II (which is imprinted) results in a mouse born small, but appears to have normal incremental growth following birth [6, 7]. That is to say that the time required for the animal to double its size approximates normal.

(4) The IGF-binding proteins (IGFBPs) act, in part, to modulate IGF action at the local/tissue level [8, 9]. In the face of targeted deletion of IGFBPs, the effects on growth are at most modest, suggesting they serve to 'fine tune' tissue-specific IGF action. However, overexpression of some IGF binding proteins, such as IGF-binding protein 1 (IGFBP-1), act as 'anti-IGF factors' and thereby restrict growth. The IGFBPs also exert effects independent of IGFs, in some cases binding at the cell surface, in others following nuclear localization.

(5) Targeted deletions of intracellular molecules involved in IGF-1R signaling generally have modest effects on growth, most likely due to pathway integration and redundancy. For example, there are multiple isoforms of AKT, a phosphorylation target of PI3 kinase that mediates IGF-I and insulin action. Targeted deletion of a single AKT isoform reduces birth weight slightly [10–14], but a combinatorial knockout of AKT and AKT isoforms yields mice born small like the IGF-1R null mutants [15].

In summary: IGF-I and IGF-II are major regulators of growth, with IGF-II perhaps playing the more important role during fetal life. The type 1 IGF receptor is a critical component in the growth control pathway and graded alterations in its abundance or signaling affect growth. The roles of the IGFBPs are complex and diverse, modulating the actions and local abundance of IGFs as well as exerting specific effects of their own.

So what is the evidence that such observations in mice are applicable to man? First, the IGF system components (ligands, receptors, etc.) are highly conserved among vertebrates, with the murine versions basically duplicated in humans. Second, low circulating concentrations of IGF-I and increases in IGFBP-1 are commonly observed in experimental models of fetal growth retardation [16–22] and in unselected groups of human infants born small for gestational age (SGA) [23–27]. Finally, recent descriptions of humans with genetic abnormalities in the IGF pathway seem to have been reasonably predicted by the observations in mice with similar genetic lesions (table 2).

Disorders of the IGF Pathway

Defects in IGF Production

In 1996, Woods et al. [28] described the first case of congenital IGF-I deficiency due to a partial deletion in the *IGF-I* gene. The patient was homozygous for the genetic lesion; parents were carriers and related. He had no detectable circulating IGF-I but normal blood levels of IGF-II and IGFBP-3. He also had severe intrauterine growth retardation (IUGR) with a birth weight of 1.4 kg at term, marked postnatal growth failure, and a number of important additional clinical features. The patient was microcephalic with mental retardation and had sensorineural deafness. Dysmorphic features included micrognathia, ptosis, low hairline, and clinodactyly. Furthermore, the patient developed carbohydrate intolerance as an adolescent. Subsequently, Bonapace et al. [29] and Wahlenkamp et al. [30] described addition cases with similar features.

That humans with congenital IGF-I deficiency would have severe pre- and postnatal growth failure was predicted by the phenotype of murine IGF-I null mutants. These cases, though seemingly rare, are particularly informative as they define the specific effects of IGF insufficiency on human fetal growth and subsequent development. Growth attenuation is the direct result of the lack of IGF-I stimulation on all growing tissues including the skeletal growth plates. Microcephaly and deafness are explained by the known requirements of IGF-I for normal brain and otological development [31, 32]. An increase in GH secretion (with direct anti-insulin effect) combined with defective insulin release due to reduced IGF-I action at the islet β-cell is postulated to cause the carbohydrate intolerance [33, 34].

Interestingly, IGF-II is a poor substitute for IGF-I as fetal growth is substantially reduced and postnatal growth very slow in the face of normal circulating IGF-II concentrations in cases of isolated IGF-I deficiency due to *IGF-I* gene mutations. Nevertheless, the important role of IGF-II in human growth is evident from the role of IGF-II deficiency in Silver-Russell syndrome [35, 36]. This disorder is characterized by severe fetal growth retardation, peculiar triangular facies, occasional hemihypertrophy of the limbs, and much reduced adult height. The *IGF-II* gene is imprinted; normally only the paternal allele is expressed. However, in approximately 50% of cases of Silver-Russell syndrome there is reduced methylation of the adjacent imprinting control region (ICR) of the paternal allele, which silences *IGF-II* gene expression [37, 38]. The reduced IGF-II abundance during fetal life thus explains the intra-uterine growth retardation. Approximately 10% of the cases with Silver-Russell syndrome show uniparental isodysomy for chromosome 7, suggesting involvement of another imprinted locus. The cause of Silver-Russell syndrome in the remaining cases has yet to be defined as well as whether IGF signaling pathways might be involved.

Table 2. Mutations causing intrauterine growth retardation in humans

Gene(s)	Disorder	Major clinical features	Comments	References
IGF-I	IUGR-short stature	severe pre- and postnatal growth failure, microcephaly, deafness, carbohydrate intolerance	based on reports of 3 families	[28–30]
IGF-II	Silver-Russell syndrome	severe pre- and postnatal growth failure	epigenetic mutation	[35, 36, 38]
IGF-1R	IUGR-short stature	severe pre-and postnatal growth failure, variable CNS abnormalities	variable increases in circulating levels of IGF-I	[30, 42, 43, 45]
Insulin	congenital diabetes	fetal growth retardation, diabetes mellitus	10% of cases of permanent neonatal DM	[55]
KCNJ11, ABCC8 (KATP channel)	congenital diabetes	fetal growth retardation, diabetes mellitus, transient or permanent	40% of cases of permanent neonatal DM CNS disease in some *KCNJ11* mutations Activating mutations	[56–58]
Chromosome 6 ICR abnormality	congenital diabetes	fetal growth retardation, transient diabetes mellitus	70% of transient neonatal DM associated with paternal isodisomy chromosome 6 or DNA methylation variation	[63]
IR	Donohue (Leprechaun) syndrome, Rabson-Mendenhall syndrome	fetal growth retardation, diabetes mellitus, moderate to severe insulin resistance	treatment with rhIGF-I may be beneficial	[61, 62]
PTF1A, IPF1	pancreatic agenesis	fetal growth retardation, diabetes mellitus	cerebellar involvement with *PTF1A*	[60]
FANC A-M	Fanconi syndrome	severe pre- and postnatal growth failure, absent thumb/radius	associated with GH deficiency, hypothyroidism, hypogonadism, and malignancy	[66, 67]
BLM	Bloom syndrome	severe pre- and postnatal growth failure	increased risk for neoplasms	[68]

Table 3. Characteristics of humans with *IGF-1R* gene mutations

Mutation	Birth weight	Adult height	IGF-I	Mental status
R108Q[1]	−2.0	−2.8	−0.3	normal
K115N[1]	−2.0	−1.6	−2.0	normal
R108Q/ K115N[1]	−3.5	−4.8	+1.0–4.0	IQ 124, OCD?
R59X[1]	−2.4	−2.6	NA	NA
R59X[1]	−3.0	NA	+1.1–2.3	delayed development
R59X[1]	−2.7	NA	NA	NA
E1050K[2]	−2.1	−4.0	+1.6	IQ 112
E1050K[2]	−3.3	NA	+2.9	normal
R709Q[3]	−1.6	−2.9	+0.5	normal
R709Q[3]	−1.5	NA	+1.5	IQ 60
R481Q[4]	−2.5	NA	'elevated'	NA
Mean ± SD	−2.4 ± 0.7	−3.1 ± 1.0		

Anthropometric measures and circulating IGF-I concentrations are expressed as standard deviation scores. Mutations shown as amino acid substitution using a single letter code.
[1]Abuzzahab et al. [42]. [2]Walenkamp et al. [44]. [3]Kawashima et al. [45]. [4]Inagaki et al. [43].

Defects in IGF Action

Defects in IGF-1R also cause fetal growth retardation in humans. This was first suspected in patients who had small deletions in chromosome 15 that encompassed the *IGF-1R* gene, resulting in haplo-insufficiency [39–41]. These infants had severe fetal growth retardation and significant neuro-cognitive deficits. In 2003, Abuzzahab et al. [42] reported the first cases of IGF resistance due to intrinsic defects in the *IGF-1R* gene. One case had a nonsense mutation resulting in haploinsufficiency, indicating that IGF resistance could explain the poor growth in the chromosome 15 deletion syndromes. The other was a compound heterozygote with missense mutations resulting in amino acid substitutions in the hormone binding region of the receptor. Since the initial description, additional cases have been reported (table 3), and begin to reveal the features common to the syndrome [43–45]. These are: a variable degree of both pre- and postnatal growth retardation, circulating IGF-I concentrations above the norms for age and gender, a range of CNS abnormalities, and dysmorphic features in some.

The significant variation in clinical features observed in those with *IGF-1R* mutations must, in part, reflect variations in the IGF signaling that result from each particular mutation. It is noteworthy that the phenotype of haploinsufficiency is severe and all of the other reported cases have perhaps less disruptive mutations. Thus far, no *IGF-1R* null lesions have been reported in humans, suggesting that a homozygous deletion may be lethal, as is the case in mice. This verifies that the IGF-1R is a critical component of human growth regulation and that graded alterations in signaling result in clinically meaningful changes in growth. It also means that the phenotypes of individuals harboring genetic variations in *IGF-1R* are likely to be wide ranging and, at times, subtle.

Clinical Features

Prenatal growth retardation without 'catch-up' growth is expected in affected patients and many have some CNS involvement (microcephaly, low IQ) as well. Carbohydrate intolerance is also found [46, 47]. Beyond this, it is difficult to hold fast to a 'classic' collection of clinical features because of the small number of cases described thus far as well as our incomplete understanding of pathophysiology of underlying defects. Indices of growth hormone secretion, such as IGFBP-3 and GH response to provocation, should be elevated because of reduced feedback at the pituitary. Indeed, this is described in several cases, but is not universal. The diagnosis of Silver-Russell syndrome is suspected when the growth deficit is accompanied by the typical facies ± hemi-hypertrophy. Serum concentrations of IGF-II are surprisingly normal [38, 48], probably reflecting the use of an alternate promoter within the liver after birth. Low circulating IGF-I in a child with a hearing deficit should lead one to suspect *IGF-I* gene mutation. A higher than average IGF-I level may be indicative of an *IGF-1R* gene defect, as circulating IGF-I concentrations are typically in the lower half of normal in the unselected short child born SGA.

It is also worthwhile to consider the effects of gene dosage observed thus far as they would have implications for the pedigree. There is perhaps some compensation for the haploinsufficiency of *IGF-I* since growth of affected mice is normal and human carriers of one mutant IGF-I allele have only a modest reduction in stature [28–30]. Gene dosage is clearly important for *IGF-II*; bi-allelic expression that occurs Beckwith-Wiedeman syndrome results in in utero overgrowth. The *IGF-1R* is perhaps the most sensitive to perturbation with apparently mild, hypomorphic mutations resulting in reduced IGF signaling and a growth retardation and haploinsufficiency producing severe growth retardation (table 3).

Diagnosis

The diagnosis is made by demonstrating biologically significant changes in *IGF-I* or *IGF-1R* gene sequences or finding increased methylation of the imprinting control regions (ICR) at the *IGF-II* gene locus. This methylation abnormality appears in approximately 50% of RSS, but has not been found in patients with nonsyndromic IUGR [49].

Management

Patients with documented *IGF-I* or *IGF-1R* gene defects should be monitored for the development of carbohydrate intolerance/diabetes mellitus. Recombinant human IGF-I (rhIGF-I) is approved for use in humans with severe primary IGF-I deficiency (defined as height and circulating IGF-I concentrations below – 3 SD for age and sex). Patients with *IGF-I* gene defects would meet these criteria and should benefit from treatment. Camacho-Hubner et al. [50] treated such a patient with 80 µg/kg/day rhIGF-I as a single injection and reported increased growth and improved insulin sensitivity. Patients with forms of IGF resistance can respond to pharmacologic GH treatment [42, 51], but the ultimate effect on height in this subset is not known and GH treatment could reduce carbohydrate tolerance.

Disorders of the Insulin Pathway

Insulin has been referred to as the 'fetal growth factor'. Humans with congenital/neonatal diabetes of all forms are characteristically small at birth and infants of diabetic mothers are large, with elevated circulating concentrations of insulin. Insulin stimulates fetal metabolism and placental nutrient transport. Thus, any condition that inhibits insulin release or action may result in fetal growth retardation. However, the notion that insulin acts directly and independently as a growth stimulator is too simplistic. Multiple lines of evidence indicate its effects are intertwined with those of the IGFs. These include a dependence on insulin for the normal production of IGF-I and regulation of IGFBP-1. Insulin suppresses IGFBP-1 expression, so when insulin secretion or action is low, IGFBP-1 is increased and sequesters the IGFs resulting in reduced IGF signaling [52, 53]. Indeed, increased IGFBP-1 is common in small-for-gestational-age human neonates and in most experimental models of IUGR [23, 25, 54].

Defects in Pancreatic Development

Genetic defects in pancreatic development that include both endocrine and exocrine portions occur rarely and lead to congenital diabetes associated with fetal growth retardation. Mutation of the pancreas transcription factor 1 (*PTF1A*) gene causes neonatal diabetes mellitus with cerebellar hypoplasia/agenesis and dysmorphism consisting of beaked nose and low set, dysplastic ears and a triangular face. Pancreatic exocrine and endocrine deficiency also occurs in mutation in the gene for insulin promoter factor-1 *(IPF1)*.

β-Cell Defects

Mutations in the insulin gene (*INS*), as might be expected, cause neonatal diabetes and influence fetal growth. Most of these are heterozygous missense mutations that likely interfere with processing/release of the protein. They are responsible for about 10% of cases of permanent neonatal diabetes [55].

Defects in genes that regulate the release of insulin have also been found. When glucose enters the β-cell, it is metabolized by glucokinase, which causes an increase in ATP and inhibition of the KATP channel which ultimately results in insulin release. Activating mutations involving the subunits of the KATP channel (SUR1 and Kir6.2) inhibit glucose-stimulated membrane depolarization and thereby attenuate insulin release [56–58]. They are found about 40% of the time. Heterozygous mutations in the glucokinase (*GCK*) gene result in maturity onset diabetes of youth (MODY). Homozygous *GCK* mutations are a rare cause of neonatal diabetes mellitus [59, 60].

Defects in Insulin Action
The insulin receptor (IR) is widely distributed among tissues. When absent, there are profound effects on muscle and fat metabolism [61, 62]. Donohue syndrome (leprechaunism) refers to severe forms of congenital insulin resistance usually due to mutations in both copies of the *IR* gene that result in an absence of the receptor. This causes severe fetal growth retardation and neonatal diabetes and is often lethal. Rabson-Mendenhall syndrome is the eponym given to intermediate forms of insulin resistance typically caused by missense mutations that reduce, but do not abrogate, insulin action. These patients typically still show relatively severe IUGR, but more modest disturbance in carbohydrate metabolism.

Clinical Features
All affected individuals show variable degrees of IUGR and carbohydrate intolerance/diabetes, depending on the nature of the lesion. Fat stores are typically depleted. Patients with insulin resistance have acanthosis nigricans and may show signs of virilization if there has been significant chronic hyperinsulinism. Some of the activating mutations in the *KCNJ11* gene (Kir6.2) are associated with a more severe phenotype that includes CNS manifestations of developmental delay and seizures.

Approximately half of the cases of neonatal diabetes are transient; the majority of these have abnormalities linked to an ICR on chromosome 6. Thus far it appears that all chromosome 6- based disease is transient. The remainder carry mutations in either the *KCNJ11* or *ABCC8* gene, which may result in permanent or transient diabetes mellitus. Patients with transient neonatal diabetes tend to present at younger ages and are a more frequently SGA at birth. However, there is substantial overlap so that it is usually impossible to predict whether the condition will remit based on clinical features [60].

Diagnosis
The precise etiology of neonatal diabetes can be determined in most cases. Measurement of insulin in the face of hyperglycemia (or assessing response to insulin administration) separates patients with defects in insulin action from those with insulin deficiency. DNA sequencing identifies defects in *IR*, *KCNJ11*, *ABCC8*, *IPF1*, etc. Most of the cases of transient neonatal diabetes appear to be due to an abnormality of imprinting on chromosome 6. Though the gene(s) involved are not yet verified, *ZAC* (a transcription factor) and *HYMAI* are candidates [63]. Uniparental disomy of chromosome 6 can be demonstrated by analyzing relevant polymorphic markers; methods to determine methylation profiles are becoming more available.

Management

Insulin replacement is the treatment of choice for most with defects in insulin production. Patients with transient forms can relapse when older, and thus ongoing surveillance is required. For the mutation in the KATP channel genes, treatment with a sulfonylurea is a theoretic option and has been used successfully in some patients [64]. Severe forms of insulin resistance can be particularly challenging, especially in the *IR* null individual. Treatment with rhIGF-I in some cases results in a substantial improvement [65]. The mechanism appears to involve stimulation of intracellular proteins common to the insulin and IGF signal transduction pathways.

Disorders with Chromosomal Instability

Fanconi anemia (FA) and Bloom syndrome are very rare, related conditions that share several clinical and physiological features. They are autosomal-recessive conditions with prenatal growth failure, cytological evidence of chromosome instability and a high rate of cancer in affected individuals. FA is caused by defect in any of at least 13 genes that make up the FA complementation group *(FANC)* [66, 67]. The gene products, many of which associate to form a complex, are involved in the repair of damaged DNA. Bloom syndrome is caused by defects in the *BLM* gene which encodes a RecQ helicase that, with the FANC proteins, comprises a large complex involved in DNA repair [68]. This functional relationship likely explains the overlap of certain features between the syndromes, though each is distinct in several ways.

Clinical Features

The classic features of Fanconi anemia are bone marrow failure associated with short stature and absent thumb and/or radius [67]. The anemia typically develops during the first 10 years of life. A third have no congenital malformations and thus may show only the growth deficit before the anemia manifests. Median survival as reported in 2003 was 24 years. The cumulative incidence of malignancy, 60% of which are hematologic, is approximately 50% by age 40 years [69].

The etiology of the poor growth is complex. An intrinsic problem in growth for all cells due to prolongation of the cell cycle could explain the pre- and postnatal growth abnormality [70]. In addition, growth hormone deficiency has been reported in up to 80% of patients and primary hypothyroidism in others, which compound the growth problem [71, 72]. Impaired glucose tolerance/diabetes and hypogonadism are additional endocrine conditions found in FA.

Children with Bloom syndrome frequently show characteristic facial findings of malar hypoplasia, prominent ears and cutaneous features which include facial telangectasia, Café-au-lait spots, and areas of hypo- and hyperpigmentation. They do not have the skeletal anomalies or bone marrow failure, but do develop malignancies, principally leukemia and lymphoma with time. Diabetes mellitus with insulin resistance has been reported. The basis for growth failure is unknown [68, 73].

Diagnosis

The diagnosis is based on cytologic tests that assess chromosome stability and/or sequencing of the gene(s) in question. When FA is suspected, additional studies to assess GH and thyroid status are warranted.

Management

As noted above, the mortality in FA is very high owing to the difficulties in managing the bone marrow failure and treating the cancers that develop. Treatment is initially supportive, but bone marrow transplantation may be necessary. Both FA and Bloom syndrome patients are unusually sensitive to some chemotherapeutic drugs.

The role that GH might play in the development of malignancy raises serious questions about GH treatment of these children. Human GH is approved for use in short children born SGA and typically prescribed in doses that exceed physiological replacement. It is this author's opinion that GH at pharmacological doses is contraindicated in Bloom S and FA. The benefits for the GH-sufficient, short SGA with high risk of malignancy do not seem to outweigh the potential (though not well-defined) risks of adding GH to the regimen. This means that short SGA children in whom there is any suspicion of one of these conditions should have appropriate testing to exclude these diagnoses before GH treatment is instituted.

The approach to the FA patient with bona fide severe GH deficiency may differ. GH deficiency results in decreased muscle mass, osteopenia, lipid abnormalities, among other metabolic disturbances. Replacing GH using doses commensurate with established requirements for GH accompanied by careful monitoring of circulating concentrations of IGF-I may result in significant clinical improvements. The risk of adding GH to the patient's regimen under these conditions may be relatively modest and perhaps not preclude treatment.

Future Directions

The characterization of single gene defects that cause IUGR has provided valuable information about the factors involved in the regulation of prenatal growth in humans. Amazingly, the etiology for most cases of congenital diabetes can be determined and is restricted to a manageable number of genetic abnormalities. The genetic mechanisms responsible for the larger group of children born SGA are much more difficult to unravel. Mutations in genes encoding post-receptor elements (e.g. AKT, IRS) have not been identified as causes of IUGR, put perhaps this is a matter of time and testing the right patients. However, the pathophysiology of IUGR is complex and involves the interaction of genes and environment (smoking, obesity, maternal malnutrition) and deviate from classical Mendelian genetics. Future studies that determine how modest genetic coding variations (e.g. polymorphisms, gene copy number variations) modify the risk of being born SGA and interact with the environment will be informative. In addition, it is evident that epigenetic modifications such as methylation, histone acetylation are playing a significant role in the 'programming' of gene expression and the subsequent development of metabolic phenotypes. The identification of epigenetic targets, the factors that lead to epigenetic variation, and the effects of those changes will be extremely important areas for investigation in the future and will lead to new understanding of the mechanisms and consequences of being born SGA.

References

1 Lupu F, Terwilliger JD, Lee K, et al: Roles of growth hormone and insulin-like growth factor 1 in mouse postnatal growth. Dev Biol 2001;229:141–162.

2 Liu J-P, Baker J, Perkins AS, et al: Mice carrying null mutations of the genes encoding insulin-like growth factor I (Igf-1) and the type 1 IGF receptor (Igf1r). Cell 1993;75:59–72.

3 Ludwig T, Eggenschwiler J, Fisher P, et al: Mouse mutants lacking the type 2 IGF receptor (IGF2R) are rescued from perinatal lethality in the Igf2 and Igf1r null backgrounds. Dev Biol 1996;177:517–535.

4 Chu CH, Tzang BS, Chen LM, et al: IGF-II/mannose-6-phosphate receptor signaling induced cell hypertrophy and atrial natriuretic peptide/BNP expression via Galphaq interaction and protein kinase C-alpha/CaMKII activation in H9c2 cardiomyoblast cells. J Endocrinol 2008;197:381–390.

5 Powell-Braxton L, Hollingshead P, Warburton C, et al: IGF-I is required for normal embryonic growth in mice. Genes Dev 1993;7:2609–2617.

6 Baker J, Liu J-P, Robertson EJ, et al: Role of insulin-like growth factors in embryonic and postnatal growth. Cell 1993;75:73–82.

7 DeChiara TM, Robertson EJ, Efstratiadis A: Parental imprinting of the mouse insulin-like growth factor II gene. Cell 1991;64:849–859.

8 Mohan S, Baylink DJ: IGF-binding proteins are multifunctional and act via IGF-dependent and -independent mechanisms. J Endocrinol 2002;175:19–31.

9 Silha JV, Murphy LJ: Insulin-like growth factor binding proteins in development. Adv Exp Med Biol 2005;567:55–89.

10 Cho H, Thorvaldsen JL, Chu Q, et al: Akt1/ PKBalpha is required for normal growth but dispensable for maintenance of glucose homeostasis in mice. J Biol Chem 2001;276:38349–38352.

11 Chen WS, Xu PZ, Gottlob K, et al: Growth retardation and increased apoptosis in mice with homozygous disruption of the Akt1 gene. Genes Dev 2001;15:2203–2208.

12 Garofalo RS, Orena SJ, Rafidi K, et al: Severe diabetes, age-dependent loss of adipose tissue, and mild growth deficiency in mice lacking Akt2/PKB beta. J Clin Invest 2003;112:197–208.

13 Easton RM, Cho H, Roovers K, et al: Role for Akt3/protein kinase Bgamma in attainment of normal brain size. Mol Cell Biol 2005;25:1869–1878.

14 Tschopp O, Yang ZZ, Brodbeck D, et al: Essential role of protein kinase B gamma (PKB gamma/Akt3) in postnatal brain development but not in glucose homeostasis. Development 2005;132:2943–2954.

15 Peng XD, Xu PZ, Chen ML, et al: Dwarfism, impaired skin development, skeletal muscle atrophy, delayed bone development, and impeded adipogenesis in mice lacking Akt1 and Akt2. Genes Dev 2003;17:1352–1365.

16 Murray MA, Dickson BA, Smith EP, et al: Epidermal growth factor (EGF) stimulates insulin-like growth factor binding protein-1 (IGFBP-1) expression in the neonatal rat. Endocrinology 1993;133:159–165.

17 Unterman T, Lascon R, Gotway MB, et al: Circulating levels of insulin-like growth factor binding protein-1 (IGFBP-1) and hepatic mRNA are increased in the small for gestational age (SGA) fetal rat. Endocrinology 1990;127:2035–2037.

18 Jones CT, Gu W, Harding JE, et al: Studies on the growth of the fetal sheep. Effect of surgical reduction of placental size or experimental manipulation of uterine blood flow on plasma sulphation promoting activity and on the concentration of insulin-like growth factors I and II. J Dev Physiol 1988;10: 179–189.

19 Unterman TG, Glick RP, Hollis RF, et al: Insulin-like growth factor I (IGF-I), insulin, and nutritional status are reduced in the small for gestational age (SGA) fetal rat. Clin Res 1990;38:324A.

20 Price WA, Stiles AD, Moats-Staats BM, et al: Gene expression of insulin-like growth factors (IGFs), and type 1 IGF receptor, and IGF-binding proteins in dexamethasone-induced fetal growth retardation. Endocrinology 1992;130:1424–1432.

21 Kampman KA, Ramsay TG, White ME: Developmental changes in serum IGF-1 and IGFBP levels and liver IGFBP-3 mRNA expression in intrauterine growth-retarded and control swine. Comp Biochem Physiol Biochem Molec Biol 1994;108:337–347.

22 Price WA, Rong L, Stiles AD, et al: Changes in IGF-I and -II, IGF binding protein, and IGF receptor transcript abundance after uterine artery ligation. Pediatr Res 1992;32:291–295.

23 Fant M, Salafia C, Baxter RC, et al: Circulating levels of IGFs and IGF binding proteins in human cord serum: relationships to intrauterine growth. Regul Pept 1993;48:29–39.

24 Lassarre C, Hardouin S, Daffos F, et al: Serum insulin-like growth factors and insulin-like growth factor binding proteins in the human fetus: relationships with growth in normal subjects and in subjects with intrauterine growth retardation. Pediatr Res 1991;29:219–225.

25 Reece EA, Wiznitzer A, Le E, et al: The relation between human fetal growth and fetal blood levels of insulin-like growth factors I and II, their binding proteins, and receptors. Obstet Gynecol 1994;84:88–95.

26 Giudice LC, De Zegher F, Gargosky SE, et al: Insulin-like growth factors and their binding proteins in the term and preterm human fetus and neonate with normal and extremes of intrauterine growth. J Clin Endocrinol Metab 1995;80:1548–1555.

27 Ostlund E, Bang P, Hagenas L, et al: Insulin-like growth factor I in fetal serum obtained by cordocentesis is correlated with intrauterine growth retardation. Hum Reprod 1997;12:840–844.

28 Woods KA, Camacho-Hubner C, Savage MO, et al: Intrauterine growth retardation and postnatal growth failure associated with deletion of the insulin-like growth factor I gene. N Engl J Med 1996;335: 1363–1367.

29 Bonapace G, Concolino D, Formicola S, et al: A novel mutation in a patient with insulin-like growth factor 1 (IGF1) deficiency. J Med Genet 2003;40:913–917.

30 Walenkamp MJ, Karperien M, Pereira AM, et al: Homozygous and heterozygous expression of a novel insulin-like growth factor-I mutation. J Clin Endocrinol Metab 2005;90:2855–2864.

31 Camarero G, Villar MA, Contreras J, et al: Cochlear abnormalities in insulin-like growth factor-1 mouse mutants. Hearing Res 2002;170:2–11.

32 Ye P, D'Ercole AJ: Insulin-like growth factor actions during development of neural stem cells and progenitors in the central nervous system. J Neurosci Res 2006;83:1–6.

33 Kulkarni RN, Holzenberger M, Shih DQ, et al: beta-cell-specific deletion of the Igf1 receptor leads to hyperinsulinemia and glucose intolerance but does not alter beta-cell mass. Nat Genet 2002;31:111–115.

34 Xuan S, Kitamura T, Nakae J, et al: Defective insulin secretion in pancreatic beta cells lacking type 1 IGF receptor. J Clin Invest 2002;110:1011–1019.

35 Abu-Amero S, Monk D, Frost J, et al: The genetic aetiology of Silver-Russell syndrome. J Med Genet 2008;45:193–199.

36 Gicquel C, Rossignol S, Cabrol S, et al: Epimutation of the telomeric imprinting center region on chromosome 11p15 in Silver-Russell syndrome. Nat Genet 2005;37:1003–1007.

37 Eggermann T, Schonherr N, Meyer E, et al: Epigenetic mutations in 11p15 in Silver-Russell syndrome are restricted to the telomeric imprinting domain. J Med Genet 2006;43:615–616.

38 Netchine I, Rossignol S, Dufourg MN, et al: 11p15 imprinting center region 1 loss of methylation is a common and specific cause of typical Russell-Silver syndrome: clinical scoring system and epigenetic-phenotypic correlations. J Clin Endocrinol Metab 2007;92:3148–3154.

39 Roback EW, Barakat AJ, Dev VG, et al: An infant with deletion of the distal long arm of chromosome 15 (q26.1→qter) and loss of insulin-like growth factor 1 receptor gene. Am J Med Genet 1991;38: 74–79.

40 Tamura T, Tohma T, Ohta T, et al: Ring chromosome 15 involving deletion of the insulin-like growth factor 1 receptor gene in a patient with features of Silver- Russell syndrome. Clin Dysmorphol 1993;2:106–113.

41 Siebler T, Lopaczynski W, Terry CL, et al: Insulin-like growth factor i receptor expression and function in fibroblasts from two patient with deletion of the distal long arm of chromosome 15. J Clin Endocrinol Metab 1995;80:3447–3457.

42 Abuzzahab MJ, Schneider A, Goddard A, et al: IGF-I receptor mutations resulting in intrauterine and postnatal growth retardation. N Engl J Med 2003; 349:2211–2222.

43 Inagaki K, Tiulpakov A, Rubtsov P, et al: A familial IGF-1 receptor mutant leads to short stature: clinical and biochemical characterization. J Clin Endocrinol Metab 2007.

44 Walenkamp MJ, van der Kamp HJ, Pereira AM, et al: A variable degree of intrauterine and postnatal growth retardation in a family with a missense mutation in the insulin-like growth factor I receptor. J Clin Endocrinol Metab 2006;91:3062–3070.

45 Kawashima Y, Kanzaki S, Yang F, et al: Mutation at cleavage site of IGF receptor in a short stature child born with intrauterine growth retardation. J Clin Endocrinol Metab 2005;90:4679–4687.

46 Sundararajan S, Banach W, Chernausek SD: Defective glucose homeostasis in a child with a genetically-acquired function defect in the IGF-I receptor. Proc Pediatr Soc Ann Meet, San Francisco, 2004.

47 Woods KA, Camacho-Hubner C, Bergman RN, et al: Effects of insulin-like growth factor I (IGF-I) therapy on body composition and insulin resistance in IGF-I gene deletion. J Clin Endocrinol Metab 2000;85:1407–1411.

48 Binder G, Seidel AK, Weber K, et al: IGF-II serum levels are normal in children with Silver-Russell syndrome who frequently carry epimutations at the IGF2 locus. J Clin Endocrinol Metab 2006;91:4709–4712.

49 Eggermann T, Meyer E, Caglayan AO, et al: ICR1 epimutations in llp15 are restricted to patients with Silver-Russell syndrome features. J Pediatr Endocrinol Metab 2008;21:59–62.

50 Camacho-Hubner C, Woods KA, Miraki-Moud F, et al: Effects of recombinant human insulin-like growth factor I (IGF-I) therapy on the growth hormone-IGF system of a patient with a partial IGF-I gene deletion. J Clin Endocrinol Metab 1999;84:1611–1616.

51 Walenkamp MJ, de Muinck Keizer-Schrama SM, de Mos M, et al: Successful long-term growth hormone therapy in a girl with haploinsufficiency of the IGF-I receptor due to a terminal 15q26.2 → qter deletion detected by multiplex ligation probe amplification. J Clin Endocrinol Metab 2008;93:2421–2425.

52 Murphy LJ: Overexpression of insulin-like growth factor binding protein-1 in transgenic mice. Pediatr Nephrol 2000;14:567–571.

53 Lewitt MS, Saunders H, Phyual JL, et al: Regulation of insulin-like growth factor-binding protein-1 in rat serum. Diabetes 1994;43:232–239.

54 Ali O, Cohen P: Insulin-like growth factors and their binding proteins in children born small for gestational age: implication for growth hormone therapy. Horm Res 2003;60(suppl 3):115–123.

55 Edghill EL, Flanagan SE, Patch AM, et al: Insulin mutation screening in 1,044 patients with diabetes: mutations in the INS gene are a common cause of neonatal diabetes but a rare cause of diabetes diagnosed in childhood or adulthood. Diabetes 2008;57:1034–1042.

56 Babenko AP, Polak M, Cave H, et al: Activating mutations in the ABCC8 gene in neonatal diabetes mellitus. N Engl J Med 2006;355:456–466.

57 Gloyn AL, Pearson ER, Antcliff JF, et al: Activating mutations in the gene encoding the ATP-sensitive potassium-channel subunit Kir6.2 and permanent neonatal diabetes. N Engl J Med 2004;350:1838–1849.

58 Gloyn AL, Siddiqui J, Ellard S: Mutations in the genes encoding the pancreatic beta-cell KATP channel subunits Kir6.2 (KCNJ11) and SUR1 (ABCC8) in diabetes mellitus and hyperinsulinism. Hum Mutat 2006;27:220–231.

59 Glaser B: Insulin mutations in diabetes: the clinical spectrum. Diabetes 2008;57:799–800.

60 Polak M, Cave H: Neonatal diabetes mellitus: a disease linked to multiple mechanisms. Orphanet J Rare Dis 2007;2:12.

61 Taylor SI, Cama A, Accili D, et al: Mutations in the insulin receptor gene. Endocr Rev 1992;13:566–595.

62 Accili D, Barbetti F, Cama A, et al: Mutations in the insulin receptor gene in patients with genetic syndromes of insulin resistance and acanthosis nigricans. J Invest Dermatol 1992;98:77S–81S.

63 Arima T, Drewell RA, Arney KL, et al: A conserved imprinting control region at the HYMAI/ZAC domain is implicated in transient neonatal diabetes mellitus. Hum Mol Genet 2001;10:1475–1483.

64 Flechtner I, Vaxillaire M, Cave H, et al: Diabetes in very young children and mutations in the insulin-secreting cell potassium channel genes: therapeutic consequences. Endocr Dev 2007;12:86–98.

65 McDonald A, Williams RM, Regan FM, et al: IGF-I treatment of insulin resistance. Eur J Endocrinology 2007;157(suppl 1):S51–S56.

66 Wang W: Emergence of a DNA-damage response network consisting of Fanconi anaemia and BRCA proteins. Nat Rev 2007;8:735–748.

67 Bagby GC, Alter BP: Fanconi anemia. Semin Hematol 2006;43:147–156.

68 Kaneko H, Kondo N: Clinical features of Bloom syndrome and function of the causative gene, BLM helicase. Expert Rev Molecular Diagn 2004;4:393–401.

69 Kutler DI, Singh B, Satagopan J, et al: A 20-year perspective on the International Fanconi Anemia Registry (IFAR). Blood 2003;101:1249–1256.

70 Akkari YM, Bateman RL, Reifsteck CA, et al: The 4N cell cycle delay in Fanconi anemia reflects growth arrest in late S phase. Mol Genet Metab 2001;74:403–412.

71 Wajnrajch MP, Gertner JM, Huma Z, et al: Evaluation of growth and hormonal status in patients referred to the International Fanconi Anemia Registry. Pediatrics 2001;107:744–754.

72 Giri N, Batista DL, Alter BP, et al: Endocrine abnormalities in patients with Fanconi anemia. J Clin Endocrinol Metab 2007;92:2624–2631.

73 Thomas ER, Shanley S, Walker L, et al: Surveillance and treatment of malignancy in Bloom syndrome. Clin Oncol 2008;20:375–379.

74 Zhou Y, Xu BC, Maheshwari HG, et al: A mammalian model for Laron syndrome produced by targeted disruption of the mouse growth hormone receptor/binding protein gene (the Laron mouse). Proc Natl Acad Sci USA 1997;94:13215–13220.

75 DeChiara TM, Efstratiadis A, Robertson EJ: A growth-deficiency phenotype in heterozygous mice carrying an insulin-like growth factor II gene disrupted by targeting. Nature 1990;345:78–80.

76 Efstratiadis A: Genetics of mouse growth. Int J Dev Biol 1998;42:955–976.

77 Araki E, Lipes MA, Patti ME, et al: Alternative pathway of insulin signalling in mice with targeted disuption of the IRS-1 gene. Nature 1994;372:186–190.

78 Tamemoto H, Kadowaki T, Tobe K, et al: Insulin resistance and growth retardation in mice lacking insulin receptor substrate-1. Nature 1994;372:182–186.

79 Withers DJ, Gutierrez JS, Towery H, et al: Disruption of IRS-2 causes type 2 diabetes in mice. Nature 1998;391:900–904.

80 Dummler B, Tschopp O, Hynx D, et al: Life with a single isoform of Akt: mice lacking Akt2 and Akt3 are viable but display impaired glucose homeostasis and growth deficiencies. Mol Cell Biol 2006;26: 8042–8051.

Prof. Steven D. Chernausek, MD
Director, Children's Medical Research Institute (CMRI), Diabetes and Metabolic Research Program
Department of Pediatrics, University of Oklahoma Health Sciences Center
1122 NE. 13th Street, Suite 1400, Oklahoma City, OK 73117 (USA)
Tel. +1 405 271 2767, Fax +1 405 271 3439, E-Mail steven-chernausek@ouhsc.edu

Kiess W, Chernausek SD, Hokken-Koelega ACS (eds): Small for Gestational Age. Causes and Consequences.
Pediatr Adolesc Med. Basel, Karger, 2009, vol 13, pp 60–72

Birth Weight and Later Risk of Type 2 Diabetes

Thomas Harder · Karen Schellong · Elke Rodekamp · Joachim W. Dudenhausen · Andreas Plagemann

Division of 'Experimental Obstetrics', Clinic of Obstetrics,
Charité – Universitätsmedizin Berlin, Campus Virchow-Klinikum, Berlin, Germany

Abstract

According to the 'small baby syndrome hypothesis', birth weight is claimed to show an inverse linear relation to later risk of type 2 diabetes. By looking into the literature 14 studies involving a total of 132,180 individuals can be identified of which, however, only seven displayed an inverse linear relation. Accordingly, meta-analysis of all published studies revealed that low birth weight is indeed associated with increased risk of type 2 diabetes (OR = 1.32, 95% CI: 1.06–1.64). However, high birth weight is associated with increased risk of type 2 diabetes to the same extent (OR = 1.27, 95% CI: 1.01–1.59). Meta-regression showed a U-shaped relation between birth weight and subsequent risk of type 2 diabetes. Taken together, these findings indicate that the relation between birth weight and risk of type 2 diabetes later in life is not linear inverse, but U-shaped. Pathophysiological mechanisms and consequences for preventive medicine are discussed.   Copyright © 2009 S. Karger AG, Basel

In 1993, Barker et al. [1] published a highly influential study on a relation between a low birth weight and increased risk of developing symptoms of the metabolic syndrome in adulthood. A number of studies confirmed this observation and the respective hypothesis [2–4]. The metabolic syndrome often precedes the development of type 2 diabetes [5, 6]. Therefore, the question arises whether low birth weight is also a risk factor for type 2 diabetes. Indeed, a variety of studies found an association between low birth weight and subsequently increased risk of type 2 diabetes [7–9]. This led to the widespread assumption, declared in a large number of narrative reviews, that clear evidence exists of an effect of low birth weight on the risk to develop type 2 diabetes later in life [10–12]. Furthermore, many authors claimed that the relation between birth weight and risk of type 2 diabetes is inverse linear, implying that high birth weight protects against type 2 diabetes. To the contrary, however, studies exist showing that high but not low birth weight is followed by increased risk of type 2 diabetes [13]. Moreover, in some studies an increased prevalence of type 2 diabetes was found in both low birth weight and high birth weight subjects [14, 15].

Since both high birth weight and intrauterine growth restriction show considerably increasing prevalences during recent years [16, 17], possible long-term consequences might have high relevance for public health. Given, on the one hand, this potential for primary prevention and, on the other, the obviously controversial observations mentioned above, an integrative synthesis on the current state of knowledge seems to be mandatory. We therefore performed a systematic review and meta-analysis on the relation between birth weight and subsequent risk of type 2 diabetes in later life.

A Systematic Review of the Literature

This systematic review was conducted according to the MOOSE group checklist for meta-analyses of observational studies [18]. We performed a literature search including the databases MEDLINE and EMBASE, using the terms birth weight, type 2 diabetes, non-insulin-dependent diabetes mellitus and NIDDM in the full-text option, without language restrictions. Furthermore, a manual search was carried out of all references cited in original studies and in all reviews identified. To be eligible for inclusion, studies had to fulfill the following criteria: (1) Original report investigating the relation between birth weight and type 2 diabetes in later life. (2) Odds ratio (OR) and 95% confidence interval (95% CI; or data to calculate them) of type 2 diabetes in at least two strata of birth weight were reported. Alternatively, an OR (95% CI) for change in risk of type 2 diabetes per unit change in birth weight had to be reported.

Forty articles were identified to be potentially relevant and were subjected to full review. 13 original studies met the inclusion criteria [7–9, 13–15, 19–25]. One consisted of two studies [13], so that 14 studies (10 cohort studies, 4 case-control studies) were included in this systematic review.

Study characteristics of included reports are displayed in table 1. Studies involved a total of 132,180 individuals, of whom 6,901 had type 2 diabetes. The first study was published in 1991, while the most recent appeared in 2003. Year of birth varied from 1911 to 1997, thereby including subjects born during eight decades. The studies were performed in 7 countries on 3 continents. Age at examination ranged from 6 to 75 years. Study size ranged from 138 to 69,526 probands. Only 2 of 14 reports [9, 13] considered gestational age at birth of study probands.

Twenty-seven studies had to be excluded during the review process [1, 30–55]. Most of them did not provide sufficient data to calculate an OR with 95% CI for type 2 diabetes according to birth weight. One study [39] was a twin study which had to be excluded since only adjusted OR and pooled data from twin pairs were reported which lack statistical independence. Two studies [37, 43] were double publications using subcohorts of studies or data already included in the meta-analysis [8, 31].

Of the 14 studies which could be included, ten [7, 9, 13(I), 13(II), 14, 15, 22–25] provided data for calculation of OR (95% CI) of type 2 diabetes in subjects with low birth weight (<2,500 g), compared to those above this cutoff value. From nine studies

Table 1. Characteristics of included studies

Reference, year	Country of origin	Design	Year of birth	Age years	Cohort size
Barker et al. [19], 2002	Finland	cohort	1924–1944	53–73	13,517
Carlsson et al. [20], 1999	Sweden	cohort	1938–1957	35–56	2,294
Curhan et al. [7], 1996	USA	cohort	1911–1946	40–75	22,846
Dyck et al. [13], 2001 (I)	Canada	case-control	1950–1984	10–45	1,728
Dyck et al. [13], 2001 (II)	Canada	case-control	1950–1984	10–45	2,264
Eriksson et al. [21], 2003	Finland	case-control	1934–1944	40	8,702
Fall et al. [22], 1998	India	cohort	1934–1953	39–60	501
Forsén et al. [23], 2000	Finland	cohort	1924–1933	64–73	7,044
Hales et al. [24], 1991	Great Britain	cohort	1920–1930	59–70	370
Lithell et al. [8], 1996	Sweden	cohort	1920–1924	60	1,093
McCance et al. [14], 1994	USA	cohort	1940–1972	20–39	1,179
Rich-Edwards et al. [9], 1999	USA	cohort	1921–1946	60	69,526
Wei et al. [15], 2003	Taiwan	cohort	1992–1997	6–18	978
Young et al. [25], 2002	Canada	cohort	not reported	<18	138

BMI = Body mass index; CC = case-control study; CO = cohort study; GDM = gestational diabetes mellitus; SES = socio-economic status. *See text.

[7, 9, 13(I), 13(II), 14, 15, 21, 23, 25], data for calculation of OR of type 2 diabetes in probands with high birth weight (>4,000 g [13(I), 13(II), 15, 21, 23, 25] or >4,500 g [7, 9, 14]), compared to those below this cutoff value could be extracted. In eight studies [7, 9, 13(I), 13(II), 14, 15, 23, 25] both low and high birth weight were reported. Three studies [8, 19, 20] did only report an OR with 95% CI of type 2 diabetes per 1,000 g linear increase in birth weight.

Table 1. (continued)

Cases with type 2 diabetes	Assessment of birth weight	Assessment of type 2 diabetes	Confounders	Trend
698	records	register of medication	year of birth, sex	linear inverse
35	questionnaires	clinical examinations	age, BMI, family history of diabetes	linear inverse
424	questionnaires	questionnaires	age, BMI, parental history of diabetes	linear inverse
846	records	register	age, sex, maternal age, parity, previous stillbirth, gestational age	linear positive
1,164	records	register	age, sex, maternal age, parity, previous stillbirth	linear positive
292	records	register of medication	not reported	linear inverse
75	records	clinical examinations	none	linear positive
471	records	register	weight at 7 years	linear inverse
27	records	clinical examinations	none	linear inverse
61	records	clinical examinations	BMI at 50 years	linear inverse
210	records	clinical examinations	age, BMI, maternal diabetes	u-shaped
2,123	questionnaires	records	age, adult BMI, maternal history of diabetes, gestational age	linear inverse*
429	register	clinical examinations	age, sex, BMI, family history of diabetes, SES, GDM	u-shaped
46	interview	records	diabetes during pregnancy, diet, smoking during pregnancy, alcohol during pregnancy, mother's pre-pregnancy BMI, breastfeeding	u-shaped

Seven studies reported the existence of a linear inverse relation between birth weight and risk of type 2 diabetes, while in three studies a linear positive association was found. In three studies, a u-shaped relation was declared. Remarkably, in one further study [9], a linear inverse relation was indicated in the abstract, while the data in the main text of the article indeed showed a u-shaped relation.

An Attempt at a Quantitative Data Synthesis: Meta-Analysis

We tried to quantitatively summarized the data obtained from the 14 studies which were identified by systematic review, using standard meta-analytic techniques. Low birth weight (<2,500 g) was associated with an increased risk of type 2 diabetes, compared to a birth weight equal to or above 2,500 g (OR: 1.32; 95% CI: 1.06–1.64; n = 10 studies; random-effects model). Fixed-effects model revealed a slightly higher pooled OR (OR: 1.49; 95% CI: 1.36–1.64). Results of the studies were significantly heterogeneous (p = 0.007). Influence analysis revealed that the study by Rich-Edwards et al. [9] largely influenced the pooled OR: Omitting this study from the data set led to a pooled estimate that was more close to 1.0 and was not significant (OR: 1.25; 95% CI: 0.97–1.59).

High birth weight (>4,000 g), compared to a birth weight equal to or below 4,000 g, was associated with increased risk of type 2 diabetes to the same extent as low birth weight. This effect was observed using the random-effects model (OR: 1.27; 95% CI: 1.01–1.59; n = 9 studies) as well as by using the fixed-effects model (OR: 1.26; 95% CI: 1.12–1.42). Again, significant heterogeneity was observed (p = 0.001).

Given these findings of increased risk of type 2 diabetes at both ends of the birth weight spectrum, we repeated the dichotomous comparisons, now using birth weight of 2,500–4,000 g as reference for all studies that gave data on both low and high birth weight. As expected, the pooled estimates for risk of type 2 diabetes after low as well as high birth weight increased, as compared to the reference category (low birth weight: 1.47; 95% CI: 1.26–1.72; high birth weight: 1.36; 95% CI: 1.07–1.73; both by random-effects model; n = 8 studies; fig. 1). The fixed effects model gave similar results (low birth weight: 1.55; 95% CI: 1.41–1.70; high birth weight: 1.34; 95% CI: 1.18–1.52; n = 8 studies).

From all 14 studies 54 estimates for specific categories of birth weight could be included in the meta-regression analysis. Visual inspection of the scatterplot revealed that the relationship between birth weight and risk of type 2 diabetes was u-shaped. Therefore, birth weight and (birth weight $\times$ birth weight) were included as independent variables. In the weighted meta-regression, these variables were significantly related to risk of type 2 diabetes (regression coefficients: birth weight: –0.0011; 95% CI: –0.0018 to –0.004; p = 0.003; (birth weight $\times$ birth weight): 1.50×10^{-7}; 95% CI: 4.43×10^{-8} to 2.55×10^{-7}; p = 0.005; fig. 2).

We used the data from dichotomous comparisons to evaluate whether the results might be influenced by publication bias. Neither for the relation between low birth weight and type 2 diabetes, nor for that between high birth weight and type 2 diabetes was evidence of publication bias found, as indicated by visual inspection of the funnel plots (not shown), and nonsignificant Begg's tests (for low birth weight: p = 0.72; for high birth weight: p = 0.92) and Egger's tests (for low birth weight: p = 0.20; for high birth weight: p = 0.84).

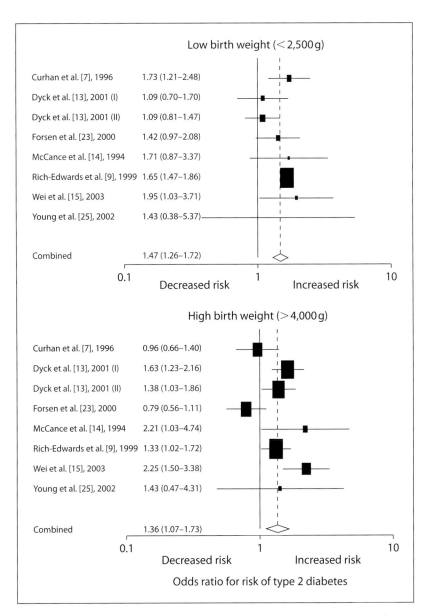

Fig. 1. Meta-analysis of the risk of type 2 diabetes in individuals with low birth weight (<2,500 g; top) or high birth weight (>4,000 g; bottom), as compared to individuals with normal birth weight (2,500–4,000 g; reference). Pooled OR were calculated by random effects model.

Discussion

Birth Weight and Type 2 Diabetes: Not an Inverse, but a U-Shaped Relation
During recent years, the issue of a relation between low birth weight and increased risk of later type 2 diabetes has raised broad interest, since potential implications for

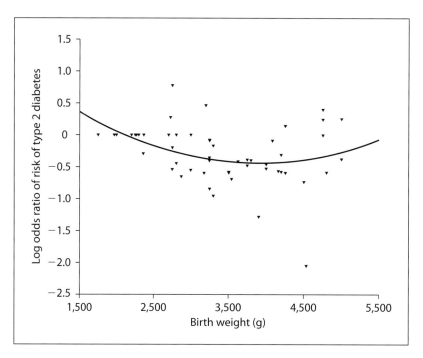

Fig. 2. Relation between birth weight and subsequent risk of type 2 diabetes in a meta-analysis of 14 studies providing 54 estimates. Weighted meta-regression revealed a u-shaped relation between birth weight and risk of type 2 diabetes (for details, see text).

preventive medicine are enormous. Meanwhile, the public perception seems to be that low birth weight is an established risk factor for type 2 diabetes. Moreover, it is often postulated that an inverse linear relation exists between birth weight and risk of type 2 diabetes, leading to the conclusion that a higher birth weight is protective and may even result in a decreased mortality from type 2 diabetes [56]. This first systematic review and meta-analysis indicates, however, that these conclusions are wrong: Applying different techniques we found, indeed, that a relation between birth weight and subsequent type 2 diabetes exists. However, it is not linear inverse, but u-shaped.

Although many more original reports were published which investigated this issue, up to now only 10 studies exist which adhered to basal standards of study quality by providing essential data to perform a quantitative data synthesis on the relation between low birth weight (<2,500 g) and risk of type 2 diabetes. Moreover, of these studies, only two [9, 13] provided estimates which are adjusted for gestational age [57]. Due to these small numbers, we could not perform subgroup analyses since the results have to be expected to be instable. Moreover, the pooled OR indicating a relation between low birth weight and risk of type 2 diabetes is largely influenced by one big study: omitting the data of Rich-Edwards et al. [9] considerably reduces the estimate towards 1.0 and non-significance, which was not the case in the analysis on high birth weight and risk of type 2 diabetes. Interestingly enough, the small number of original studies providing sufficient data on the relation between low birth weight

Harder · Schellong · Rodekamp · Dudenhausen · Plagemann

and later risk of type 2 diabetes stand in sharp contrast to the amount of 47 narrative reviews (according to MEDLINE) which declared to summarize the current knowledge on this topic. Remarkably, 46 of these 47 reviews concluded that an inverse linear relation exists between birth weight and later risk of type 2 diabetes.

In contrast, our systematic review of the published literature and meta-analysis shows that also high birth weight (>4,000 g) is a risk factor for subsequent type 2 diabetes, with the same effect size as that of low birth weight. This result was confirmed by meta-regression, indicating the existence of a u-shaped relation between birth weight and risk of type 2 diabetes. These results underscore one of the advantages of meta-analyses, since the study estimates summed up did not only came from studies that described positive linear or u-shaped relations, but also from studies which did not primarily investigate and discuss the effect of high birth weight but provided data for its calculation. Consequently, the pooled OR for risk of type 2 diabetes increased for both low and high birth weight when a 'normal' range of birth weight (2,500–4,000 g) was used as reference. This further highlights the importance of the definition of the reference group, based on an uncommitted working hypothesis, for the interpretation of results from studies on the impact of birth weight on later risk of disease. Moreover, meta-analyses which only pool linear trend data to investigate associations between birth weight and later risk of disease, without carefully checking the appropriateness of assuming linearity for the given research question, must lead to biased conclusions.

Methodological Considerations

Meta-analyses on the association between (low) birth weight and blood pressure [58] and cholesterol levels in later life [59] revealed the existence of a considerable degree of publication bias. Here, this was not the case: neither inspection of the funnel plot, nor formal tests for funnel plot asymmetry gave a statistically significant indication of publication bias. Remarkably, two of the excluded studies were double publications of subcohorts, again raising the question of publication bias, which, however, could not be confirmed statistically.

Eleven of the 14 studies of our analysis provided adjusted estimates to assess a possible role of confounding. However, the adjusted ORs were calculated by using different reference categories of birth weight, using different confounders. Moreover, most original studies suffer from a lack of sufficient information to calculate an adjusted pooled OR. Therefore, a conclusion on the impact of confounding factors on the relation between birth weight and type 2 diabetes cannot be evidently provided from this systematic review and meta-analysis.

According to basal requirements for quantitative data synthesis, 27 studies had to be excluded from this analysis, mainly because of insufficient data provided in the articles to calculate an OR. Surprisingly, among them were some of the most often cited studies on the relation between low birth weight and type 2 diabetes [1, 48]. Furthermore, a number of excluded studies only analyzed surrogate measures like insulin resistance as outcome parameters [36, 55].

A Pathophysiological Approach

The etiopathological mechanisms by which birth weight might be related to later risk of type 2 diabetes are still a matter of debate. While Barker and co-workers have claimed that the relation between low birth weight and later risk of type 2 diabetes reflects long-term consequences of in utero undernutrition [1, 60], we and others [61–64] have favored a role of specific maternal diseases during pregnancy. Worthy to note, Hofman et al. [65] provided data clearly speaking against intrauterine undernutrition as causal factor for the 'small baby syndrome', by demonstrating that term small-for-gestational-age infants are as insulin resistant later in life as preterm appropriate-for-gestational-age children. Most importantly, however, these and other data speak in favor of a role of the neonatal environment for the association between low birth weight and subsequent type 2 diabetes [57]. In particular, low birth weight babies are highly likely to be subjected to forced neonatal feeding and overfeeding, leading to rapid neonatal weight gain. Rapid neonatal weight gain, however, is dose-dependently positively related to overweight in adulthood [67, 68], i.e. the main risk factor of the metabolic syndrome and type 2 diabetes. Moreover, these epidemiologic observations are supported by data from animal experiments which indicate that neonatal overnutrition leads to rapid neonatal weight gain, followed by overweight and diabetogenic disturbances in later life, due to 'malprogramming' of neuroendocrine circuits regulating appetite control, body weight, and metabolism [63, 64, 69, 70].

The association between high birth weight and increased risk of type 2 diabetes may reflect, at least in part, perinatal malprogramming due to exposure to undiagnosed and nontreated maternal hyperglycemia during pregnancy. Given the high prevalence of overweight women in industrialized countries, this may frequently lead to increased birth weight, causally linked to maternofetal hyperglycemia and consecutive fetal hyperinsulinism [71]. Epidemiologic and experimental studies have shown that offspring of mothers with diabetes during pregnancy have an increased risk of developing type 2 diabetes in later life [72–81].

Conclusions

Taken together, the results of this systematic review and meta-analysis show that birth weight is related in a u-shaped manner to later risk of type 2 diabetes. Data revealed that high birth weight is associated with increased risk of type 2 diabetes in later life to the same extent as low birth weight. Although it cannot be excluded that low birth weight per se may predispose to later type 2 diabetes, we would like to suggest that rather neonatal overfeeding, which is likely to occur in low birth weight babies, might be an etiological key factor standing behind the association between low birth weight and later diabetes risk. Moreover, in the case of high birth weight it appears probable that fetal overnutrition is followed by neonatal overnutrition as well, since women giving birth to high birth weight babies are highly likely to have gestational diabetes

Harder · Schellong · Rodekamp · Dudenhausen · Plagemann

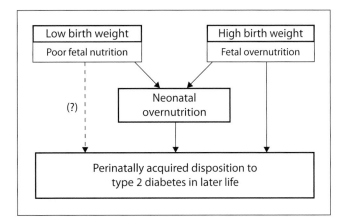

Fig. 3. Basic pathophysiological concept of the relation between low or high birth weight and subsequent risk of type 2 diabetes.

and/or are obese, thereby strongly increasing the probability that their offspring are exposed to formula instead of breast milk or to an unhealthy lifestyle during lactation which might negatively affect breast milk composition (fig. 3). This implies enormous chances for future strategies of the primary prevention of type 2 diabetes. However, while there is no longer any doubt on the 'programming' impact of perinatal conditions, for the development of causal strategies for primary prevention research is especially needed to uncover etiopathogenic mechanisms standing behind these associations.

References

1 Barker DJP, Hales CN, Fall CHD, Osmond C, Phipps K, Clark PMS: Type 2 (non-insulin-dependent) diabetes mellitus, hypertension and hyperlipidaemia (syndrome X): relation to reduced fetal growth. Diabetologia 1993;36:62–67.

2 Barker DJP, Martyn CN, Osmond C, Hales CN, Fall CHD: Growth in utero and serum cholesterol concentrations in adult life. Br Med J 1993;307:1524–1527.

3 Phillips DIW, Hirst S, Clark PMS, Hales CN, Osmond C: Fetal growth and insulin secretion in adult life. Diabetologia 1994;37:592–596.

4 Leon DA, Lithell HO, Vagero D, Koupilova I, Mohsen R, Berglund L, Lithell UB, McKeigue PM: Reduced fetal growth rate and increased risk of death from ischaemic heart disease: cohort study of 15,000 Swedish men and women born 1915–29. Br Med J 1998;317:241–245.

5 Reaven GM: Role of insulin resistance in human disease. Banting Lecture 1988. Diabetes 1988;37:1595–1607.

6 Ford ES, Giles WH, Mokdad AH: Increasing prevalence of the metabolic syndrome among US adults. Diabetes Care 2004;27:2444–2449.

7 Curhan GC, Willett WC, Rimm EB, Spiegelman D, Ascherio AL, Stampfer MJ: Birth weight and adult hypertension, diabetes mellitus, and obesity in US men. Circulation 1996;94:3246–3250.

8 Lithell HO, McKeigue PM, Berglund L, Mohsen R, Lithell UB, Leon DA: Relation of size at birth to non-insulin dependent diabetes and insulin concentrations in men aged 50–60 years. Br Med J 1996;312:406–410.

9 Rich-Edwards JW, Colditz GA, Stampfer MJ, Willett WC, Gillman MW, Hennekens CH, Speizer FE, Manson JE: Birthweight and the risk for type 2 diabetes in adult women. Ann Intern Med 1999;130:278–284.

10 Hales CN: Non-insulin-dependent diabetes mellitus. Br Med Bull 1997;53:109–122.

11 Phillips DIW: Birth weight and the future development of diabetes: a review of the evidence. Diabetes Care 1998;21(suppl 2):B150–B155.

12 Godfrey KM, Barker DJP: Fetal nutrition and adult disease. Am J Clin Nutr 2000;71(suppl):1344S–1352S.

13 Dyck RF, Klomp H, Tan L: From 'thrifty genotype' to 'hefty fetal phenotype': the relationship between high birthweight and diabetes in Sasketchewan registered Indians. Can J Publ Health 2001;92:340–344.

14 McCance DR, Pettitt DJ, Hanson RL, Jacobsson LT, Knowler WC, Bennett PH: Birth weight and non-insulin dependent diabetes: thrifty genotype, or surviving small baby genotype. Br Med J 1994;308:942–945.

15 Wei JN, Sung FC, Li CY, Chang CH, Lin RS, Lin CC, Chiang CC, Chuang LM: Low birth weight and high birth weight infants are both at an increased risk to have type 2 diabetes among schoolchildren in Taiwan. Diabetes Care 2003;26:343–348.

16 Martin JA, Hamilton BE, Ventura SJ, Menacker F, Park MM: Births: final data for 2000. Natl Vital Stat Rep 2002;50:1–101.

17 Rooth G: Increase in birthweight: a unique biological event and an obstetrical problem. Eur J Obstet Gynecol Reprod Biol 2003;106:86–87.

18 Stroup DF, Berlin JA, Morton SC: Meta-analysis of observational studies in epidemiology: a proposal for reporting. JAMA 2000;283:2008–2012.

19 Barker DJP, Eriksson JG , Forsén T, Osmond C: Fetal origins of adult disease: strength of effects and biological bias. Int J Epidemiol 2002;31:1235–1239.

20 Carlsson S, Persson PG, Alvarsson M, Efendic S, Norman A, Svanström L, Ostenson CG, Grill V: Low birth weight, family history of diabetes, and glucose intolerance in Swedish middle-aged men. Diabetes Care 1999;22:1043–1047.

21 Eriksson JG, Forsén T, Tuomilehto J, Osmond C, Barker DJ: Early adiposity rebound in childhood and risk of type 2 diabetes in adult life. Diabetologia 2003;46:190–194.

22 Fall CHD, Stein CE, Kumaran K, Cox V, Osmond C, Barker DJ, Hales CN: Size at birth, maternal weight, and type 2 diabetes in South India. Diabet Med 1998;15:220–227.

23 Forsén T, Eriksson JG, Tuomilehto J, Reunanen A, Osmond C, Barker DJ: The fetal and childhood growth of persons who develop type 2 diabetes. Ann Intern Med 2000;133:176–182.

24 Hales CN, Barker DJP, Clark PMS, Cox LJ, Fall C, Osmond C, Winter PD: Fetal and infant growth and impaired glucose tolerance at age 64. Br Med J 1991;303:1019–1022.

25 Young TK, Martens PJ, Taback SP, Sellers EA, Dean HJ, Cheang M, Flett B: Type 2 diabetes mellitus in children: prenatal and early infancy risk factors among native Canadians. Arch Pediatr Adolesc Med 2002;156:651–655.

26 Guyer B, Hoyert DL, Martin JA, Ventura SJ, MacDorman MF, Strobino DM: Annual summary of vital statistics – 1998. Pediatrics 1999;104:1229–1246.

27 Naylor CD, Sermer M, Chen E, Sykora K: Cesarean delivery in relation to birth weight and gestational glucose tolerance: pathophysiology or practice style? Toronto Trihospital Gestational Diabetes Investigators. JAMA 1996;275:1165–1170.

28 Thompson SG, Sharp SJ: Explaining heterogeneity in meta-analysis: a comparison of methods. Stat Med 1999;18:2693–2708.

29 Greenland S, Longnecker MP: Methods for trend estimation from summarized dose-response data, with applications to meta-analysis. Am J Epidemiol 1992;135:1301–1309.

30 Amini J, Han AM, Beracochea E, Bukenya G, Vince JD: Anthropometrical antecedents of non-insulin dependent diabetes mellitus: an age and sex matched comparison study of anthropometric indices in schoolchildren from a high prevalence Port Moresby community. Diabetes Res Clin Pract 1997;35:75–80.

31 Bhargava SK, Sachdev HS, Fall CHD, Osmond C, Lakshmy R, Barker DJ, Biswas SK, Ramji S, Prabhakaran D, Reddy KS: Relation of serial changes in childhood body-mass index to impaired glucose tolerance in young adulthood. N Engl J Med 2004;350:865–875.

32 Burke JP, Forsgren J, Palumbo PJ, Bailey KR, Desai J, Devlin H, Leibson CL: Association of birth weight and type 2 diabetes in Rochester, Minnesota. Diabetes Care 2004;27:2512–2513.

33 Cook JTE, Levy JC, Page RC, Shaw JA, Hattersley AT, Turner RC: Association of low birth weight with ß-cell function in the adult first degree relatives of non-insulin-dependent diabetic subjects. Br Med J 1993;306:302–306.

34 Dabelea D, Hanson RL, Bennett PH, Roumain J, Knowler WC, Pettitt DJ: Increasing prevalence of type 2 diabetes in American Indian children. Diabetologia 1998;41:904–910.

35 Eriksson JG, Forsén T, Tuomilehto J, Osmond C, Barker DJ: Fetal and childhood growth and hypertension in adult life. Hypertension 2000;36:790–794.

36 Eriksson JG, Forsén T, Tuomilehto J, Jaddoe VWV, Osmond C, Barker DJP: Effects of size at birth and childhood growth on the insulin resistance syndrome in elderly individuals. Diabetologia 2002;45:342–348.

37 Eriksson JG, Forsén T, Osmond C, Barker DJ: Pathways of infant and childhood growth that lead to type 2 diabetes. Diabetes Care 2003;26:3006–3010.

38 Hyppönen E, Power C, Smith GD: Prenatal growth, BMI, and risk of type 2 diabetes by early midlife. Diabetes Care 2003;26:2512–2517.

39 Iliadou A, Cnattingius S, Lichtenstein P: Low birth weight and type 2 diabetes: a study on 11,162 Swedish twins. Int J Epidemiol 2004;33:948–953.

40 Kubaszek A, Markkanen A, Eriksson JG, Forsen T, Osmond C, Barker DJ, Laakso M: The association of the K121Q polymorphism of the plasma cell glycoprotein-1 gene with type 2 diabetes and hypertension depends on size at birth. J Clin Endocrinol Metab 2004;89:2044–2047.

41 Lindsay RS, Dabelea D, Roumain J, Hanson RL, Bennett PH, Knowler WC: Type 2 diabetes and low birth weight: the role of paternal inheritance in the association of low birth weight and diabetes. Diabetes 2000;49:445–449.

42 Lindsay RS, Hanson RL, Wiedrich C, Knowler WC, Bennett PH, Baier LJ: The insulin gene variable number tandem repeat class I/III polymorphism is in linkage disequilibrium with birth weight but not type 2 diabetes in the Pima population. Diabetes 2003;52:187–193.

43 McKeigue PM, Lithell HO, Leon DA: Glucose tolerance and resistance to insulin-stimulated glucose uptake in men aged 70 years in relation to size at birth. Diabetologia 1998;41:1133–1138.

44 Mi J, Law CM, Zhang K: Association of body size at birth with impaired glucose tolerance during their adulthood for men and women aged 41 to 47 years in Beijing of China. Zhonghua Yu Fang Yi Xue Za Zhi 1999;33:209–213.

45 Mitchell SMS, Weedon MN, Owen KR, Shields B, Wilkins-Wall B, Walker M, McCarthy MI, Frayling TM, Hattersley AT: Genetic variation in the small heterodimer partner gene and young-onset type 2 diabetes, obesity, and birth weight in UK subjects. Diabetes 2003;52:1276–1279.

46 Nelson RG, Morgenstern H, Bennett PH: Birth weight and renal disease in Pima Indians with type 2 diabetes mellitus. Am J Epidemiol 1998;148:650–656.

47 Pettitt DJ, Knowler WC: Long-term effects of the intrauterine environment, birth weight, and breast-feeding in Pima Indians. Diabetes Care 1998;21 (suppl 2):138–141.

48 Phipps K, Barker DJP, Hales CN, Fall CH, Osmond C, Clark PM: Fetal growth and impaired glucose tolerance in men and women. Diabetologia 1993;36: 225–228.

49 Poulsen P, Vaag AA, Kyvik KO, Moller Jensen D, Beck-Nielsen H: Low birth weight is associated with NIDDM in discordant monozygotic and dizygotic twin pairs. Diabetologia 1997;40:439–446.

50 Rasmussen SK, Lautier C, Hansen L, Echwald SM, Hansen T, Ekstrom CT, Urhammer SA, Borch-Johnsen K, Grigorescu F, Smith RJ, Pedersen O: Studies of the variability of the genes encoding the insulin-like growth factor I receptor and its ligand in relation to type 2 diabetes mellitus. J Clin Endocrinol Metab 2000;85:1606–1610.

51 Shield JPH, Gardner RJ, Wadsworth EJK, Whiteford ML, James RS, Robinson DO, Baum JD, Temple IK: Aetiopathology and genetic basis of neonatal diabetes. Arch Dis Child Fetal Neonatal Ed 1997;67: F39–F42.

52 Smith GD, Sterne JAC, Tynelius P, Rasmussen F: Birth characteristics of offspring and parental diabetes: evidence for the fetal insulin hypothesis. J Epidemiol Community Health 2004;58:126–128.

53 Stanner SA, Bulmer K, Andrès C, Lantseva OE, Borodina V, Poteen VV, Yudkin JS: Does malnutrition in utero determine diabetes and coronary heart disease in adulthood? Results from the Leningrad siege study: a cross sectional study. Br Med J 1997; 315:1342–1348.

54 Steinhart JR, Sugarman JR, Connell FA: Gestational diabetes is a herald of NIDDM in Navajo women: high rate of abnormal glucose tolerance after GDM. Diabetes Care 1997;20:943–947.

55 Valdez R, Athens MA, Thompson GH, Bradshaw BS, Stern MP: Birthweight and adult health outcomes in a biethnic population in the USA. Diabetologia 1994;37:624–631.

56 Syddall HE, Sayer AA, Simmonds SJ, Osmond C, Cox V, Dennisson EM, Barker DJ, Cooper C: Birth weight, infant weight gain, and cause-specific mortality: the Herfordshire Cohort Study. Am J Epidemiol 2005;161:1074–1080.

57 Plagemann A, Harder T: Premature birth and insulin resistance (letter). N Engl J Med 2005;352: 939–940.

58 Huxley R, Neil A, Collins R: Unravelling the fetal origins hypothesis: is there really an inverse association between birth weight and subsequent blood pressure? Lancet 2002;360:659–665.

59 Huxley R, Owen CG, Whincup PH, Cook DG, Colman S, Collins R: Birth weight and subsequent cholesterol levels. JAMA 2004;292:2755–2767.

60 Hales CN, Ozanne SE: For Debate: Fetal and early postnatal growth restriction lead to diabetes, the metabolic syndrome and renal failure. Diabetologia 2003;46:1013–1019.

61 Morley R, Owens J, Blair E, Dwyer T: Is birthweight a good marker for gestational exposures that increase the risk of adult disease? Paediatr Perinat Epidemiol 2002;16:194–199.

62 Harder T, Kohlhoff R, Dörner G, Rohde W, Plage-mann A: Perinatal 'programming' of insulin resistance in childhood: critical impact of neonatal insulin and low birth weight in a risk population. Diabetic Med 2001;18:634–639.

63 Plagemann A, Rodekamp E, Harder T: To: Hales CN, Ozanne SE: For debate: Fetal and early postnatal growth restriction lead to diabetes, the metabolic syndrome and renal failure. Diabetologia 2003;46:1013–1019 (letter). Diabetologia 2004;47:1334–1335.

64 Plagemann A: 'Fetal programming' and 'functional teratogenesis': on epigenetic mechanisms and prevention of perinatally acquired lasting health risks. J Perinat Med 2004;32:297–305.

65 Hofman PL, Regan F, Jackson WE, Jefferies C, Knight DB, Robinson EM, Cutfield WS: Premature birth and later insulin resistance. N Engl J Med 2004;351:2179–2186.

66 Bazaes RA, Mericq V: Premature birth and insulin resistance (letter). N Engl J Med 2005;359:939–940.

67 Stettler N, Stallings VA, Troxel AB, Zhao J, Schinnar R, Nelson SE, Ziegler EE, Strom BL: Weight gain in the first week of life and overweight in adulthood: a cohort study of European American subjects fed infant formula. Circulation 2005;111:1897–1903.

68 Euser AM, Finken MJ, Keijzer-Veen MG, Hille ET, Wit JM, Dekker FW, Dutch POPS-19 Collaborative Study Group: Associations between prenatal and infancy weight gain and BMI, fat mass, and fat distribution in young adulthood: a prospective cohort study in males and females born very preterm. Am J Clin Nutr 2005;81:480–487.

69 Plagemann A, Harder T, Rake A, Voits M, Fink H, Rohde W, Dörner G: Perinatal increase of hypothalamic insulin, acquired malformation of hypothalamic galaninergic neurons, and syndrome X-like alterations in adulthood of neonatally overfed rats. Brain Res 1999;836:146–155.

70 Boullu-Ciocca S, Dutour A, Guillaume V, Achard V, Oliver C, Grino M: Postnatal diet-induced obesity in rats upregulates systemic and adipose tissue glucocorticoid metabolism during development and in adulthood. Diabetes 2005;54:197–203.

71 Weiss PAM: Gestational diabetes: a survey and the Graz approach to diagnosis and therapy; in Weiss PAM, Coustan DR (eds): Gestational Diabetes. Wien, Springer, 1988, pp 1–58.

72 Pettitt DJ, Baird HR, Aleck KA, Bennett PH, Knowler WC: Excessive obesity in offspring of Pima Indian women with diabetes during pregnancy. N Engl J Med 1983;308:242–245.

73 Pettitt DJ, Bennett PH, Knowler WC, Baird HR, Aleck KA: Gestational diabetes mellitus and impaired glucose tolerance during pregnancy: long-term effects on obesity and glucose tolerance in the offspring. Diabetes 1985;34(suppl 2):119–122.

74 Dabelea D, Hanson RL, Lindsay RS, Pettitt DJ, Imperatore G, Gabir MM, Roumain J, Bennett PH, Knowler WC: Intrauterine exposure to diabetes conveys risks for type 2 diabetes and obesity: a study of discordant sibships. Diabetes 2000;49:2208–2211.

75 Silverman BL, Rizzo T, Green OC, Cho NH, Winter RJ, Ogata ES, Richards GE, Metzger BE: Long-term prospective evaluation of offspring of diabetic mothers. Diabetes 1991;40(suppl 2):121–125.

76 Silverman BL, Purdy LS, Metzger BE: The intrauterine environment: implications for the offspring of diabetic mothers. Diab Rev 1996;4:21–35.

77 Plagemann A, Harder T, Kohlhoff R, Rohde W, Dörner G: Glucose tolerance and insulin secretion in children of mothers with pregestational insulin-dependent diabetes mellitus or gestational diabetes. Diabetologia 1997;40:1094–1100.

78 Aerts L, Van Assche FA: Is gestational diabetes an acquired condition? J Dev Physiol 1979;2:19–25.

79 Aerts L, Holemans K, Van Assche FA: Maternal diabetes during pregnancy: consequences for the offspring. Diabetes Metab Rev 1990;6:147–167.

80 Dörner G, Plagemann A: Perinatal hyperinsulinism as possible predisposing factor for diabetes mellitus, obesity and enhanced cardiovascular risk in later life. Horm Metab Res 1994;26:213–221.

81 Plagemann A, Harder T, Melchior K, Rake A, Rohde W, Dörner G: Elevation of hypothalamic neuropeptide Y-neurons in adult offspring of diabetic mother rats. Neuroreport 1999;10:3211–3216.

Prof. Dr. med. Andreas Plagemann, MD
Division of 'Experimental Obstetrics', Clinic of Obstetrics
Charité – Universitätsmedizin Berlin, Campus Virchow-Klinikum
Augustenburger Platz 1, DE–13353 Berlin (Germany)
Tel. +49 30 450 524 041, Fax +49 30 450 524 928, E-Mail andreas.plagemann@charite.de

Kiess W, Chernausek SD, Hokken-Koelega ACS (eds): Small for Gestational Age. Causes and Consequences.
Pediatr Adolesc Med. Basel, Karger, 2009, vol 13, pp 73–85

Low Birth Weight and Optimal Fetal Development: A Global Perspective

Kathryn L. Franko[a] · Peter D. Gluckman[a] ·
Catherine M. Law[b] · Alan S. Beedle[a] · Susan M.B. Morton[a]

[a]Liggins Institute and National Research Centre for Growth and Development,
University of Auckland, Auckland, New Zealand; [b]Centre for Paediatric Epidemiology and Biostatistics,
Institute of Child Health, University College London, London, UK

Abstract

About 20 million children are born each year with low birth weight, overwhelmingly in low income countries. The consequences emerge initially as infant and child mortality, predominantly from infectious disease, are later manifested as impaired growth and cognitive function, and are perpetuated by an intergenerational cycle of maternal underweight or overweight leading to suboptimal fetal development. In regions undergoing the nutrition transition, being born small and being exposed to high caloric intake later in life leads to metabolic mismatch, playing a major role in the epidemic of obesity and its complications. The concept of low birth weight has many limitations: it conflates maturation and growth, uses an absolute cut-off that may not be appropriate in all settings, and assumes a single standard of fetal growth. Because of these limitations, focus has recently shifted to optimizing fetal development rather than simply seeking to increase birth weight. Public health interventions aim to make the external environment optimal for the mother to nurture her fetus, whereas clinical interventions are focused on optimizing the status of the individual woman as the environment in which the fetus develops. Key to both is the recognition that infant and childhood health is dependent on maternal health.

Most children who are born small are born in low income, or developing, countries and children who are born small are more likely to die at birth or in the neonatal period, to suffer perinatal injury, to have a poor neonatal course, to have reduced infant survival, to suffer from infections, to have cognitive impairment, to be stunted in their physical growth and, as this chapter discusses, to have consequences that may appear later in the life course or even in the next generation.

The term 'low birth weight' is generally used to define infants with a weight below 2,500 g at the time of delivery, and this cut-off has been used extensively in population comparisons and public health surveillance in both low and high income, or developed, countries. Nevertheless, use of a single definition to identify children at risk may not be appropriate in all contexts.

In this review of suboptimal fetal development in low-income countries, we first briefly review the incidence of low birth weight in the developing world. We then consider the appropriateness of the conventional definition of low birth weight in this setting, and present a case for a life course perspective on fetal development rather than a focus on birth weight alone. We then review the causes and consequences of poor fetal development, and end by outlining a strategy aimed at improving child and adult health throughout life by optimizing fetal development.

The Global Burden of Low Birth Weight

Of the approximately 130 million children born each year in the world, about 20 million are born at a low birth weight [1]. These children are born overwhelmingly in low income countries [2], representing 16% of births there [3], and over half of them are born in South Asia [1]. The incidence of low birth weight in developing countries has been reported to range from less than 5% in China and Polynesia, through nearly 30% in India, to 50% in Bangladesh [3]. As nearly 60% of infants are not weighed at delivery in low income countries even when the birth is managed by a skilled attendant [1], these numbers may be underestimated.

The consequences of low birth weight emerge initially as infant and child mortality. Of the 4 million neonatal deaths annually, 60–80% are in low birth weight babies [4]. Low birth weight in developing countries tracks into underweight during childhood [5], contributing to the 50% of children in South Asia who are underweight [6], and 53% of the 10 million deaths each year of children under the age of 5 years are attributed to being underweight [2]. The immediate cause of death in a high proportion of such children is infectious disease, related to the impaired immune status conferred by being underweight [2]. Early nutritional effects on immunocompetence are likely to persist into adulthood, as demonstrated by the strong relationship between seasonality of birth and mortality from infections among young adults in a population of subsistence farmers in the Gambia [7].

The high incidence of low birth weight in the developing world is in part caused by chronic undernutrition of the mother before pregnancy. In South Asia, 60% of women are underweight; this is likely to represent a reflection of their own small size at birth and its later consequences and to contribute to a recurring intergenerational cycle of impaired development [8].

About 15% of the total global burden of all disease can be attributed to the joint effects of childhood and maternal underweight or micronutrient deficiencies [9]. Despite this significant impact, few attempts have been made to model the economic benefits of interventions to reduce the incidence of low birth weight in developing countries. A single study has estimated a lifetime benefit of USD 510 per low birth weight event avoided [10]; although methodological limitations to that study mean that this value is almost certainly an underestimate, it is telling for the economic development of low-income countries that over half of that benefit arose from productivity gains. Further evidence for influences of early development on human capital comes from a cohort study of US adults affected in utero by the influenza pandemic of 1918 and who demonstrated significantly impaired educational and socioeconomic attainment as adults [11].

Low Birth Weight: A Flawed Indicator?

The concept of low birth weight has many limitations: it assumes that maturation and growth can be combined in a single measure, that the concept of an absolute cut-off (2,500 g) is logical and appropriate in all contexts, and that single standards of fetal growth are meaningful. The use of low birth weight as an absolute indicator of health status also generates the danger that decreasing the rate of low birth weight becomes a goal of intervention rather than an (imperfect) indicator of the broader aim of improving maternal and infant health. The limitation of directing intervention efforts at the 'problem' of low birth weight is that birth weight is, at best, only an indirect and partial measure of fetal health and pregnancy outcome [12]. It is important to emphasize that birth size is simply a currently useful and reasonably accessible index of development to date and of the potential life course trajectory that will follow.

Importantly, low birth weight does not distinguish between suboptimal fetal growth and shortened gestation, although the two have different risk factors and are amenable to different interventions [13]. Gestation length is shorter in populations where nutritional and health status is low, as in the Netherlands after the Dutch Famine of 1944–1945 [14] and in rural India [Fall CHD, pers. commun.]. There is growing experimental [15] and epidemiological [16] evidence that the nutritional status of mothers at conception can influence gestational length. This creates a major challenge for public health, particularly in low income countries, as it is no longer appropriate to focus interventions solely on child-bearing women, but now the focus must become all women of child-bearing age. Even a mild reduction in gestation length, which in a developed country may be of little significance, can in the context of the developing world have severe impacts on the later health of that infant. In Canada, the recent fall in perinatal mortality is independent of any change in mean birth weight, and has occurred despite a rising prevalence of low birth weight [17]. In such countries, trends in rates of low birth weight may not reflect trends in perinatal health, at least as measured by early mortality [18].

Birth weight is also context-specific, making comparisons between populations problematic. For example, birth weight is influenced by parity, and in some populations such in most of China [19] and western Europe [20], the majority of infants are first-born. Additionally, the factors that determine differences in birth weight within populations, and the extent to which they are amenable to intervention, may differ from those that operate between populations [21].

The problems of low birth weight and small size for gestational age are of particular significance in the developing world. Here the limitations of the use of the concept of an absolute cut-off weight are clear. In India, the mean birth weight is less than 2,700 g, only marginally above the recognized low birth weight cut-off. This distribution is unlikely to be driven exclusively by genetic components, but probably results in part from low maternal stature and subsequently increased maternal constraint. Whether or not these maternal conditions reflect solely behavioural conditions including generations of under-nutrition, poor health and overwork, or also a genetic component limiting stature, remains uncertain.

An existing conundrum is how to define optimal birth size. An infant of a given birth weight within the 'normal range' might be appropriate in size for the context of its pregnancy or might be relatively adipose due to, among other causes, maternal gestational diabetes, or it might have been destined to be larger and in fact be growth impaired, possibly with a reduced ponderal index. For example, a 3,000 g baby born at term in India might be in any one of these three categories depending on maternal health, stature and the environment of the pregnancy. A research priority is to find measurable biomarkers for infants that might indicate a suboptimal fetal environment even if birth size is within the normative range.

The Concept of Optimal Fetal Development: A Life Course Perspective

It is not possible to link causally low birth weight and later outcomes; rather, low birth weight is an index of the overall status of the mother-infant dyad. Early interpretations of the developmental origins of adult disease paradigm tended to this view – namely that there was a distinct outcome causally related to being born small and only later was it recognized that birth size in this situation is a surrogate measure reflecting a less than optimal fetal environment.

When fetal growth is severely impaired, it may well reflect on a specific fetal pathology (such as a chromosomal disorder). But most often fetal growth is impaired by interference with the supply of nutrients to the fetus, either because of maternal or placental compromise or both. It is important in this context to recognize that maternal non-genomic factors constrain fetal growth [22] and this constraint is greater in some pregnancies (for example, when mothers are shorter, in first pregnancy, or very young). In this review, we will not focus on the specific causes of intrauterine growth retardation, as they are covered elsewhere in this volume. Rather, we shall focus on the globally dominant cluster of disadvantage and disempowerment which has adverse affects on the fetus and its life course. Infants with impaired fetal growth and maturation are often manifestations of this cycle of disadvantage. Their mothers are often of low socioeconomic status and more likely to have had less than optimal nutrition, stunted stature, and to have received little or no education. These factors increase the likelihood that low birth weight infants will grow into children and eventually adults who are also of low socioeconomic status, undernourished and uneducated. Throughout their lives, their health is at risk because of their poor start to life.

Because of the limitations of a focus on birth weight, international health agencies have recently begun to discuss optimizing fetal development rather than simply seeking to increase birth weight as a strategy to promote health and productivity [23]. Optimal fetal development describes a set of circumstances that leads to healthy mothers bearing infants with expectations of long and healthy lives. Suboptimal fetal development therefore implies that the current and/or future health of at least one component of the materno-fetal dyad is compromised. The conditions a fetus experiences during pregnancy are influenced by a wide range of factors, including but not limited to the conditions its mother experienced in utero, the health of its mother before and during pregnancy (which determines the nutritional, physical and emotional environment to which the fetus is exposed), the length of gestation and birth size. Rather than viewing birth as an end point for fetal development or a starting point for infant development, it is merely a single event in the developmental continuum that starts with maternal conditions before conception and extends into adulthood. This life course perspective has the advantage of allowing progress in developmental biology and the understanding of developmental plasticity [24] to be better integrated with advances in epidemiology and for these diverse disciplines to be utilized together to advance population health.

Causes of Suboptimal Fetal Development

'No fact is better established than that the death rate, and especially the death rate among children, is high and in inverse proportion to the social status of the population.'

Newsholme, 1910 [25]

The association between poverty and high infant morbidity and mortality has been documented for nearly 200 years [26, 27]. Recent international intervention efforts have shifted focus from the deprived child to the deprived mother [28], as the intrauterine hormonal and nutritional milieu not only influences neonatal and child health, but susceptibility to disease in adulthood as well [29, 30]. Poverty in its broadest sense encompasses physical, psychological and social deprivation, all of which are known contributors to poor fetal outcome.

Individuals in low income countries often have limited access to health care and education, and hence pregnant women are less likely to receive prenatal care, or to take advantage of it if made available, contributing to poor health status before and during pregnancy. The most common maternal conditions contributing to poor fetal outcome are generally treatable or preventable and include excessive manual labor, unbalanced or impaired diet, anemia, diarrhea, food insecurity, domestic violence, and infectious diseases including malaria [29]. Poverty and disempowerment are central to this cycle of disadvantage. In general, until women are empowered and educated, it is difficult to see how this cycle of disadvantage will be broken in many societies. Measures of educational achievement in women are the best predictors of birth size across nations [31]. It is imperative to facilitate access to health systems and knowledge of how and when to utilize such resources so that women recognize health problems and seek out treatment and care. Creating such circumstances requires a provision of resources and infrastructure to build and supply these solutions, and training within a population to utilize them effectively.

Unbalanced nutrition in pregnant women increases their susceptibility to disease and impairs their ability to produce healthy offspring. The specific nutritional demands of pregnancy are poorly defined, and general nutritional recommendations depend on maternal age, parity, and health at conception. Teenage pregnancy is the norm in many developing countries, and most teenage mothers are primiparous. First-born children are smaller than their subsequent siblings, a normal circumstance that is compounded by nutrient competition between the fetus and its young, still growing, mother [32]. Infants born to teenage mothers consistently have higher mortality rates than those born to women aged 20 years and above [33]. The inherent problems of nutrition associated with teenage pregnancy are further compounded by the limitations that childbearing places directly on maternal education access and employment opportunity and indirectly on financial security and then health care access.

Contrary to popular assumptions, undernutrition is not simply a problem of calorie deficiency. Recent data suggest that overweight and underweight adults often coexist within the same households, populations and countries. This co-occurrence suggests that malnourishment of pregnant women within those groups is still a widespread concern, even in areas where obesity and associated health problems are emerging crises [29]. Malnutrition can coexist with a state of food security for several reasons. Despite the availability of calories, protein and nutrient-rich foods may be scarce or expensive; when they are available, pregnant women may not have direct access to them because of prioritization of other household members, or may abdicate any claim to them in favor of their already-born children. Ignorance about what to consume during pregnancy may be prevalent in education-deprived populations, thus compounding the problem of nutritional deficit [29]. Furthermore, signs of malnutrition are often invisible to the untrained eye [29] so individuals do not seek out extra resources until circumstances become dire.

When pregnant women are malnourished, it is highly likely that after birth their already-deficient offspring will be malnourished in the same way. Of particular importance are iodine and vitamin A as deficiencies in each have far reaching consequences for pre-and postnatal survival, health and productivity. Globally, 35% of all people lack sufficient iodine and 40% lack adequate vitamin A in their diets [34]. Iodine deficiency in women increases the risk of spontaneous abortion, and if pregnancy is maintained, such deficiency is associated with increased rates of still-

Franko · Gluckman · Law · Beedle · Morton

birth, perinatal death and mental retardation [35]. A meta-analysis completed by the WHO reports an average loss of 13.5 IQ points after iodine deficiency in late pregnancy and early life [36], thereby also reducing human capital.

Vitamin A deficiency has equally dire consequences. Most of the developing world has moderate, severe or clinical vitamin A deficiency, and the pregnant women within those populations are very likely to be among the deficient [37]. Any available maternal stores are usurped during pregnancy and lactation, especially in the last trimester [37], and lack of this essential nutrient compromises immunity, making pregnant women and their children more susceptible to the preventable, treatable infectious diseases that already plague them due to limited access to healthcare [38]. Infants and children who are vitamin A deficient have a 23% greater chance of dying from infectious disease and diarrhea [38]. Up to 500,000 children go blind each year as a direct result of vitamin A deficiency and, of those, half will die within 1 year of becoming blind [37]. Even when food is available of adequate quantity and quality to meet the metabolic demands of pregnancy, malnutrition can still occur due in the immediate postnatal period in the absence of exclusive breastfeeding, due to poor sanitation and contaminated water. Such conditions increase neonatal mortality by a factor of 7 [39].

Stress [40, 41], workload and air pollution also contribute negatively to pregnancy outcomes. In Mexico, women who worked more than 50 h per week, stood for more than 7 h per day, or had no antenatal work-leave were at increased risk of premature delivery [42]. Similarly, low birth weight was positively associated with increased maternal workload in women in Pakistan [43] and Guatemala [44]. Air pollution is also associated with increasing numbers of low birth weight deliveries in China [45], the Czech Republic [46] and South Korea [47].

Consequences of Suboptimal Fetal Development

Infant mortality rates are significantly higher for low birth weight than non-low birth weight infants [4]. Perinatal deaths, most of which are due to suboptimal fetal development, represent at least 40% of all deaths under 5 years of age and, excluding stillbirth, are in excess of 4 million per annum globally [4]. Population studies reveal that infants born small are four times more likely to die as neonates (within 28 days of delivery) and twice as likely to die in the post-neonatal period than are their larger, heavier and healthier peers [48].

There is considerable evidence that being born small is associated with impaired cognitive development in early childhood [49] and young adulthood [50]. Poor health during pregnancy is associated with a greater risk of perinatal asphyxia and its gross neurocognitive outcomes, but there is growing evidence that the cerebral hemispheric volume of growth retarded and/or premature infants is smaller [51] and growth retarded children have higher incidences of special education needs, attention deficit disorders and poor cognitive achievement [52, 53]. In the experimental animal, growth impairment is associated with deficits in attention and learning. In both human and economic models, the impact of poor cognitive development is particularly costly to society [10].

In developed countries most children who are born small show catch-up in linear growth, although this was not always so [54, 55]. Stunting remains a common problem for children born small in least developed countries [8], and stunted children are more likely to be ill and compete less well in emergent economies [10]. On the other hand, children either born small or large are more likely to develop obesity – the different pathways by which this occurs are reviewed elsewhere [56, 57]. Extensive epidemiological data, supported by recent animal research, shows that individuals exposed to a suboptimal intrauterine environment are more prone to noncommunicable diseases as well, particularly cardiovascular and metabolic diseases [58]. A particular issue in the developing world is the rapid nature of the nutritional transition in which the intake of high glycemic index and fatty foods is rising rapidly. This is often associated with a migration to cities and a reduction in physical workloads. The combination of being born smaller and being exposed to high caloric intakes in infancy and childhood leads to metabolic mismatch and plays a major role in the rising epidemic of obesity and its complications in the developing world [59].

Developing a Strategy to Optimize Fetal Development

Given the influence of fetal development on lifetime and intergenerational health and human capital, there is an increasing focus on developing a life course approach to create an integrated strategy to optimize fetal development. Such an approach recognizes two distinct but related concepts. First, the ability of a mother to ensure optimal fetal development of her unborn baby is a reflection of accumulated health experience throughout her own development. Second, early interventions are likely to have more profound and lasting impacts than interventions delivered later in the life course, and over time will lead to improvement in health over successive generations.

Several periods of the life course should be considered: the childhood and adolescence of the girl destined to become a mother, the immediate pre-pregnancy and periconceptional period, early pregnancy, established pregnancy, the peripartum period, the neonatal period, and beyond. Over a longer time base, intergenerational factors must also be considered. Some interventions, such as a successful initiative to stop smoking in pregnancy, will have an almost immediate impact, while others will take much longer to be realized – for example, any measure that obviates infant and childhood stunting will improve fetal development in the next generation.

Table 1. Strength of evidence for interventions to optimize fetal development

Myriad evidence	Moderate evidence	Limited evidence
Decreased adolescent pregnancies	Decreased occupational hazards	Increased social support
Improved nutrition in women and girls	Safe drinking water	Increased self-esteem
Decreased infectious and other diseases	Improved maternal education	Increased sleep duration
Decreased lifestyle risk factors (workload, smoking, alcohol)	Decreased pregnancy complications	Decreased number of sexual partners, increased duration of relationship, cohabitation
Decreased environmental toxins		Decreased late pregnancy intercourse

This table is derived from the report of the WHO consultation [23]. It details some potential interventions in pregnancy which might improve outcomes for mother or child. They are classified according to the considered strength of evidence.

Given the wide range of influences on fetal development, any improvement strategy must be equally broad ranging and the challenge is to develop an affordable but holistic approach. Many public health interventions aim to make the environment optimal for the mother to nurture her fetus. Clinical interventions are focused on optimizing the status of the individual woman as the environment in which the fetus develops. A strategic approach will use both types of interventions. Table 1 shows some possible components of a strategy, arranged according to an assessment of the strength of the evidence to support them. A key element is the recognition that infant and childhood health is dependent on the health of the mother. Therefore, strategies must consider the overall status of the materno-fetal dyad. A major focus must be on the education and empowerment of girls and women of reproductive age so that they can choose when to conceive and are in appropriate health and nutritional status at conception. Such a focus should also lead to other benefits for women, their families and their communities. A specific concern is to ensure that efforts to optimize fetal development build on strategies that secure appropriate health care, such as having a trained attendant at every delivery. Basic health care is a major factor influencing maternal and child survival. Compromising the former puts the latter at considerable risk.

More specific measures include avoidance of smoking, drugs, alcohol and heavy workloads, and encouraging exclusive breastfeeding for 6 months to reduce the incidence of neonatal and infant morbidity and mortality. Many of these measures could be achieved by building on and changing focus within existing health care systems. A major research gap is the absence of detailed information on optimal maternal nutrition. Implementation of and participation in nutrient supplementation programs to increase micronutrient intake while concurrently increasing availability of adequate macronutrients will radically change the intrauterine milieu for developing fetuses and infants. However, data on the best macronutrient balance required to optimize fetal development is less certain and any recommendations must be general, although the evidence is compelling for the effects of some micronutrients, as mentioned previously [29, 60]. Furthermore, the ability of a woman to make best use of available nutrients will depend on factors unrelated to diet and food supply. Reduction of infection and decreasing workload can lead to significant improvements in nutritional status for the same dietary intake [61].

All interventions, both specific and general, need to be considered in the cultural, socioeconomic and biological context in which they will be delivered. For example, reducing the number of teenage women giving birth will have a significant impact on improving fetal development in some countries, but the specific approaches needed to reduce the incidence of adolescent pregnancy will vary in different cultures and settings, depending on the prevalence of young motherhood and the social and cultural background in which it occurs.

Conclusions

There is sufficient evidence and consensus to focus on developing a strategy aimed at optimizing fetal development rather than growth alone. This paradigm shift requires an appreciation of the importance of the maternofetal dyad and a multidisciplinary life-course approach encompassing survival, morbidity and development. Developing a strategy to optimize fetal development challenges well-established professional and lay views, but a focus on the entire course of development rather than a single point in it is likely to have broad, far-reaching benefits for pregnancy outcome and, eventually, child and adult health. While single nutrient supplementations are functional tools, they will not provide a 'magic bullet' solution to malnutrition during pregnancy and early life. Access to material and social resources, including health care and education, will directly and indirectly elevate the socioeconomic position of those most in need, thus increasing human survival and increasing human capital and quality of life. Any improvement in fetal development will have a substantial impact on long-term global health.

References

1 United Nations Childrens Fund (UNICEF), World Health Organization: Low birthweight: country, regional and global estimates, 2004. Available online: who.int/reproductive-health/publications/low_birthweight/low_birthweight_estimates.pdf (accessed June 28, 2007).

2 Black RE, Morris SS, Bryce J: Child survival. I. Where and why are 10 million children dying every year? Lancet 2003;361:2226–2234.

3 de Onis M, Blossner M, Villar J: Levels and patterns of intrauterine growth retardation in developing countries. Eur J Clin Nutr 1998;52:S5–S15.

4 Lawn J, Cousens S, Zupan J, Lancet Neonatal Survival Steering Team: 4 million neonatal deaths: When? Where? Why? Lancet 2005;365:891–900.

5 Mason JB, Hunt J, Parker D, Jonsson U: Investing in child nutrition in Asia. Asian Dev Rev 1999;17:1–32.

6 Ramalingaswami V, Jonsson U, Rohde J: The South Asian enigma; in: The Progress of Nations. New York, UNICEF, 1996, pp 10–17.

7 Moore SE, Cole TJ, Collinson AC, Poskitt EME, McGregor IA, Prentice AM: Prenatal or early postnatal events predict infectious deaths in young adulthood in rural Africa. Int J Epidemiol 1999;28:1088–1095.

8 Allen LH, Gillespie SR, United Nations Administrative Committee on Coordination, Subcommittee on Nutrition (ACC/SCN): What Works? A Review of the Efficacy and Effectiveness of Nutrition Interventions. Nutrition Policy Paper 19. Geneva, ACC/SCN in collaboration with Asian Development Bank, 2001.

9 Ezzati M, Lopez AD, Rodgers A, Vander Hoorn S, Murray CJ, Comparative Risk Assessment Collaborating Group: Selected major risk factors and global and regional burden of disease. Lancet 2002;360:1347–1360.

10 Alderman H, Behrman JR: Reducing the incidence of low birth weight in low-income countries has substantial economic benefits. World Bank Res Observer 2006;21:25–48.

11 Almond D: Is the 1918 influenza pandemic over? Long-term effects of in utero influenza exposure in the post-1940 US population. J Polit Econ 2006;114:672–712.

12 Harding JE: The nutritional basis of the fetal origins of adult disease. Int J Epidemiol 2001;30:15–23.

13 Kramer MS: The epidemiology of adverse pregnancy outcomes: an overview. J Nutr 2003;133:1592S–1596S.

14 Susser M, Stein Z: Timing in prenatal nutrition: a reprise of the Dutch Famine Study. Nutr Rev 1994;52:84–94.

15 Oliver MH, Jacquiery AL, Bloomfield FH, et al: The effects of maternal nutrition around the time of conception on the health of the offspring. Soc Reprod Fertil 2007;64(suppl):397–410.

16 Morton SMB: Maternal nutrition and fetal growth and development; in Gluckman PD, Hanson MA (eds): Developmental Origins of Health and Disease. Cambridge, Cambridge University Press, 2006, pp 98–129.

17 Joseph KS, Kramer MS, Allen AC, Cyr M, Fair M, Ohlsson A, Wen SW: Gestational age- and birthweight-specific declines in infant mortality in Canada, 1985–94. Paediatr Perinat Epidemiol 2000;14:332–339.

18 Parker JD, Schoendorf KC, Kiely JL: A comparison of recent trends in infant mortality among twins and singletons. Paediatr Perinat Epidemiol 2001;15:12–18.

19 Hesketh T, Zhu WX: Health in China. The one child family policy: the good, the bad, and the ugly. BMJ 1997;314:1685.

20 World Health Organization: Statistics by Country or Region: Total Fertility Rate, 2003. Available online: www.who.int/whosis/country/compare.cfm?language-english&country = deu&indicator = strTFR2003 (accessed March 28, 2006).

21 Wilcox AJ: On the importance – and the unimportance – of birthweight. Int J Epidemiol 2001;30:1233–1241.

22 Gluckman PD, Hanson MA: Maternal constraint of fetal growth and its consequences. Semin Fetal Neonatal Med 2004;9:419–425.

23 World Health Organization: Promoting Optimal Fetal Development: Report of a Technical Consultation. Geneva, WHO, 2006. Available online: http://www.who.int/nutrition/topics/fetal_dev_report_EN.pdf [Accessed 2007 July 19].

24 Gluckman PD, Hanson MA, Beedle AS: Early life events and their consequences for later disease: a life history and evolutionary perspective. Am J Hum Biol 2007;19:1–19.

25 Newsholme A: 39th Annual Report of the Local Government Board (England). Report Cd 5312. London, Local Government Board, 1910.

26 Vuorinen HS: Social variation in infant mortality in a core city of Finland during the 19th and early 20th centuries. Scand J Soc Med 1991;19:248–255.

27 Barker DJP: Mothers, Babies and Health in Later Life, ed 2. Edinburgh, Churchill-Livingstone, 1998.

28 Pathmanathan I, Liljestrand J, Martins JM, et al: Investing in Maternal Health: Learning from Malaysia and Sri Lanka. Washington, World Bank Publications, 2003. Available online: www-wds. worldbank.org/external/default/WDSContentServe r/WDSP/IB/2003/05/30/000094946_030516040506 82/Rendered/PDF/multi0page.pdf (accessed July 4, 2007).

29 World Bank: Repositioning Nutrition as Central to Development: A Strategy for Large-Scale Action. Washington, World Bank, 2006. Available online: http://siteresources.worldbank.org/NUTRITION/ Resources/281846–1131636806329/NutritionStrategy. pdf (accessed July 19, 2007).

30 Gluckman PD, Hanson MA: The consequences of being born small: an adaptive perspective. Horm Res 2006;65(suppl 3):5–14.

31 O'Connor KC, Morton SMB, Gluckman PD: The macroeconomic and social correlates of a poor start to life: an international comparison of low birth weights (abstract). DOHaD Conference Proceedings, 2007, submitted.

32 Naeye RL: Teenaged and pre-teenaged pregnancies: consequences of the fetal-maternal competition for nutrients. Pediatrics 1981;67:146–150.

33 Alam N: Teenage motherhood and infant mortality in Bangladesh: maternal age-dependent effect of parity one. J Biosoc Sci 2000;32:229–236.

34 World Bank: Enriching Lives: Overcoming Vitamin and Mineral Malnutrition in Developing Countries. Washington, World Bank, 1994.

35 United Nations Children's Fund (UNICEF): Statistics: Iodine Deficiency Disorders. Available online: www.childinfo.org/areas/idd/ (accessed July 19, 2007).

36 Bleichrodt N, Born MP: A meta-analysis of research on iodine and its relationship to cognitive development; in Stanbury JB (ed): The Damaged Brain of Iodine Deficiency. New York, Cognizant Communication, 1994, pp 195–200.

37 World Health Organization: Micronutrient Deficiencies: Vitamin A Deficiency. Available online: www. who.int/nutrition/topics/vad/en/ (accessed July 19, 2007).

38 United Nations Children's Fund (UNICEF): Statistics: Vitamin A Deficiency. Available online: http:// childinfo.org/areas/vitamina/ (accessed July 19, 2007).

39 Victora CG, Smith PG, Vaughan JP, et al: Infant feeding and deaths due to diarrhea: a case-control study. Am J Epidemiol 1989;129:1032–1041.

40 Wadhwa PD, Sandman CA, Porto M, et al: The association between prenatal stress and infant birth weight and gestational age at birth: a prospective investigation. Am J Obstet Gynecol 1993;169: 858–865.

41 Rondo PHC, Ferreira RF, Nogueira F, et al: Maternal psychological stress and distress as predictors of low birth weight, prematurity and intrauterine growth retardation. Eur J Clin Nutr 2003;57:266–272.

42 Ceron-Mireles P, Harlow S, Sanchez-Carrillo C: The risk of prematurity and small-for-gestational-age birth in Mexico City: the effects of working conditions and antenatal leave. Am J Publ Health 1996; 86:825–831.

43 Bhutta Z: Risk factors for maternal and foetal malnutrition in rural Singh; in: Final Report – National Workshop on Strategies to Address LBW in Pakistan. Karachi, Khan University and UNICEF, 2005.

44 Launer LJ, Villar J, Kestler E, et al: The effect of maternal work on fetal growth and duration of pregnancy: a prospective study. Br J Obstet Gynecol 1990;97:62–70.

45 Wang X, Ding H, Ryan L, et al: Association between air pollution and low birth weight: a community based study. Environ Health Perspect 1997;105: 514–520.

46 Bobak M, Dejmek J, Solansky I: Unfavourable birth outcomes of the Roma women in the Czech Republic and the potential explanations: a population-based study. BMC Publ Health 2005;5:106.

47 Ha EH, Hong YC, Lee BE, et al: Is air pollution a risk factor for low birth weight in Seoul? Epidemiology 2001;12:643–648.

48 Ashworth A: Effects of intrauterine growth retardation on mortality and morbidity in infants and young children. Eur J Clin Nutr 1998;52:S34–S41.

49 Lee H, Barratt MS: Cognitive development of preterm low birth weight children at 5 to 8 years old. J Dev Behav Pediatr 1993;14:242–249.

50 Sorensen HT, Sabroe TS, Olsen J, et al: Birth weight and cognitive function in young adult life: historical cohort study. BMJ 1997;315:401–403.

51 Nosarti C, Al-Asady MHS, Frangou S, et al: Adolescents who were born very preterm have decreased brain volumes. Brain 2002;125:1616–1623.

52 Peterson BS, Vohr B, Staib LH, et al: Regional brain volume abnormalities and long-term cognitive outcome in preterm infants. JAMA 2000;284:1939–1947.

53 Peterson BS, Anderson AW, Ehrenkranz R: Regional brain volumes and their later neurodevelopmental correlates in term and preterm infants. Pediatrics 2003;111:939–948.

54 Albertsson-Wikland K, Karlberg J: Natural growth in children born small for gestational age with and without catch-up growth. Acta Pediatr 1994;399 (suppl):64–70.

55 Hokken-Koelega AC, De Ridder MA, Lemmen RJ, et al: Children born small for gestational age: do they catch up? Pediatr Res 1995;38:267–271.

56 Gluckman PD, Hanson MA: Developmental plasticity and human disease: research directions. J Int Med 2007;261:461–471.

57 Kuzawa CW, Gluckman PD, Hanson MA: Developmental perspectives on the origin of obesity; in Fantuzzi G, Mazzone T (eds): Adipose Tissue and Adipokines in Health and Disease. Totowa, Humana Press, 2007, pp 207–219.

58 Gluckman PD, Hanson MA (eds): Developmental Origins of Health and Disease. Cambridge, Cambridge University Press, 2006.

59 Gluckman PD, Hanson MA, Morton SMB, Pinal CS: Life-long echoes: a critical analysis of the developmental origins of adult disease model. Biol Neonate 2005;87:127–139.

60 Cao, XY, Jiang XM, Dou ZH, Rakeman MA, Zhang ML, O'Donnell K: Timing of vulnerability of the brain to iodine deficiency in endemic cretinism. N Engl J Med 1994;331:1739–1744.

61 Podja J, Kelly L: United Nations Administrative Committee on Coordination, Subcommittee on Nutrition (ACC/SCN): Low Birthweight: A Report Based on the International Low Birthweight Symposium, Dhaka, Bangladesh, 1999. Nutrition Policy Paper 18. New York, ACC/SCN, 2000.

Dr. Kathryn L. Franko
Liggins Institute
University of Auckland, Private Bag 92019
Auckland 1023 (New Zealand)
Tel. +64 9 373 7599 (ext. 83450), Fax +64 9 373 7497, E-Mail k.franko@auckland.ac.nz

Kiess W, Chernausek SD, Hokken-Koelega ACS (eds): Small for Gestational Age. Causes and Consequences.
Pediatr Adolesc Med. Basel, Karger, 2009, vol 13, pp 86–98

Insulin-Like Growth Factor/Growth Hormone Axis in Intrauterine Growth and Its Role in Intrauterine Growth Retardation

W. Kiess[a] · J. Kratzsch[b] · M. Knüpfer[a] · E. Robel-Tillig[a] ·
F. Pulzer[a] · R. Pfaeffle[a]

[a]Hospital for Children and Adolescents and [b]Institute for Clinical Chemistry,
Laboratory and Molecular Medicine, University of Leipzig, Leipzig, Germany

Abstract

For medical, ethical, socioeconomic and humanitarian reasons, it seems mandatory to foster research into the causes and consequences of intrauterine growth retardation in the human. About 5% of newborns are born small for gestational age (SGA) and 10–15% of them do not naturally catch up on growth by 2 years of age. The growth of the fetus from conception to birth results from complex interactions of maternal and fetal genes with the environment, and factors such as malnutrition, infections and exposures to toxins are well known to influence fetal growth. Specific genetic disorders such as Leprechaunism, Bloom syndrome and Fanconi anemia are inherited, but are very rare causes of intrauterine growth retardation. Fetal growth depends upon substrate and oxygen supply, vascularisation of placental and fetal tissues, and endocrine modulation of cellular proliferation and tissue expansion. IGF-I has been shown to be a key stimulus of placental substrate uptake. IGF-I inhibits fetal placental catabolism and reduces placental lactate production. IGF-I deletions cause intrauterine failure to thrive. Recent published research on the actions of IGF-I in humans and the phenotypes of children with genetic defects in the GH/IGF axis establish IGF-I signaling via its receptoras the critical growth-controlling element during fetal life in man. This chapter focuses on the GH/IGF-I receptor cascade and its effect on fetal growth and its disturbance in the human.

Copyright © 2009 S. Karger AG, Basel

In a large prospective cohort study of 38,033 pregnancies in the United States, Bukowski et al. [1] asked whether or not first-trimester fetal growth was associated with birth weight, duration of pregnancy, and the risk of delivering a small-for-gestational-age (SGA) infant. 976 women from the original cohort who conceived as the result of assisted reproductive technology had a first-trimester ultrasound measurement of fetal crown-to-rump length, and delivered live singleton infants without evidence of chromosomal or congenital abnormalities. For each 1-day increase in the observed size of the fetus, birth weight increased by 28.2 g. The risk of delivering a SGA infant decreased with increasing size in the first trimester. In conclusion, variation in birth weight may be determined, at least in part, by fetal growth in the first 12 weeks after conception through effects on timing of delivery and fetal growth velocity. In a recent study from Norway, a strong interaction on intellectual performance between birth size and gestational age has been described [2].

Survival rates have greatly improved in recent years for infants of borderline viability; however, these infants remain at risk of developing a wide array of complications, not only in the neonatal unit, but also in the long term. Morbidity is inversely related to gestational age; however, there is no gestational age, including term, that is wholly exempt. Neurodevelopmental disabilities and recurrent health problems take a toll in early childhood. Subsequently, hidden disabilities such as school difficulties and behavioral problems become apparent and persist into adolescence. Reassuringly, however, most children born very preterm adjust remarkably well during their transition into adulthood. Because mortality rates have fallen, the focus for perinatal interventions is to develop strategies to reduce long-term morbidity, especially the prevention of brain injury and abnormal brain development. In addition, follow-up to middle age and beyond is warranted to identify the risks, especially for cardiovascular and metabolic disorders that are likely to be experienced by preterm survivors [3].

For medical, ethical, socioeconomic and humanitarian reasons, it seems mandatory to foster research into causes and consequences of intrauterine growth retardation in the human. This chapter deals with the putative effects of the growth hormone(GH)/insulin-like growth factor-I (IGF-I)/IGF-I receptor cascade on fetal growth and its disturbance.

Genetic and Environmental Causes of Intrauterine Growth Retardation

About 5% of newborns are SGA and 10–15% of them do not naturally catch up on growth by 2 years of age. The growth of the fetus from conception to birth results from complex interactions of maternal and fetal genes with the environment, and factors such as malnutrition, infections and exposures to toxins are well known to influence fetal growth. Specific genetic disorders such as Leprechaunism, Bloom syndrome and Fanconi anemia are inherited, but are very rare causes of intrauterine growth retardation (IUGR). Recent published research on the actions of IGF-I in humans and the phenotypes of children with genetic defects in the GH/IGF axis establish IGF-I signaling via its receptor (IGF-IR) as the critical growth-controlling element during fetal life in man [4]. The aim of other chapters in this volume is to review certain SGA disorders of Mendelian genetic origin, with an emphasis on defects in the insulin and IGF pathways which may be implicated in the persistence of short stature in some children born SGA. This chapter focuses on the

Table 1. Members of the GH-IGF axis which seem to play a pivotal role in intrauterine growth regulation

GH receptor
GH binding protein
STAT signalling molecules
Insulin-like growth factor-1
Insulin-like growth factor-2
Insulin-like growth factor-1 receptor
Insulin-like growth factor-2/mannose-6-phosphate receptor
Insulin-like growth factor binding protein-1
Insulin-like growth factor binding protein-3

Mutations and/or polymorphisms of the respective genes are thought to be linked to intrauterine growth retardation.

GH/IGF-I receptor cascade and its effect on fetal growth and its disturbance in the human.

Growth Hormone, Growth Hormone-Binding Protein and Growth Hormone Receptor Signaling

High GH bioactivity might explain the rapid growth rate of neonates. Thus, serum GH biological potency (bio-/immuno-GH ratio) and their effects on serum growth factors was studied during the first month of life in term and preterm babies [5]. Blood samples were collected from 10 small-for-gestational-age preterm (SGAPT), 17 appropriate-for-gestational age preterm (AGAPT) and 26 AGA term (T) neonates on days 4, 15 and 30 of life to evaluate serum GH values measured by IFMA (IFMA-GH) and bioassay (Bio-GH), serum IGF-I and IGF-binding protein-3 (IGFBP-3). High serum Bio-GH values on the first few days of life corresponded to high IFMA-GH values, suggesting full biological potency of circulating GH. Furthermore, IGF-I/IGFBP-3 molar ratio values in preterm babies were higher than in full-term infants. These data suggest that the high growth velocity in the first month of life of preterm neonates is due to an increased bioavailability of IGF-I [5].

Audi et al. [6] studied the frequencies of d3/fl-GHR polymorphism genotypes in control and short SGA populations. An adult control population with heights normally distributed (ACPNH) between −2 and +2 SD score (SDS) and a short non-GH-deficient SGA child population were selected. Thirty Spanish hospitals participated in the selection of the short non-GH-deficient SGA children in the setting of a controlled, randomized trial, and one of these hospitals selected the adult control population. Two hundred and eighty-nine adult subjects of both sexes constituted the controls and 247 children and adolescents of both sexes the short SGA patients. In short SGA patients, d3/fl-GHR genotype frequencies were significantly different from those in controls, with a higher frequency of the fl/fl genotype ($p < 0.0001$). In the normal controls, a trend toward diminished d3/d3 genotype frequency was observed in the shortest height group (height $<$ or $= -1$ SDS and $>$ or $= -2$ SDS, n $= 60$). It was concluded that there were in fact significant differences in the frequency distribution of the d3/fl-GHR genotypes between a normally distributed adult height population and short SGA children, with the biologically less active fl/fl genotype being almost twice as frequent in SGA patients. These data suggest that the d3/fl-GHR polymorphism might be considered among the factors that contribute to the phenotypic expression of growth [6].

As stated above, a common polymorphism in the GHR gene has been linked to increased growth response in GH-treated patients. The aim of the study by Jensen et al. [7] was to evaluate the association between the d3-GHR isoforms and spontaneous pre- and postnatal growth in a prospective study on third-trimester fetal growth velocity (FGV), birth weight, birth length, and postnatal growth. A total of 115 healthy adolescents were divided into those born SGA and appropriate for gestational age with or without intrauterine growth restriction. FGV was measured by serial ultrasonography, birth weight, birth length, and adolescent height. Isoforms of the d3-GHR gene (fl/fl, d3/fl, and d3/d3) were determined. The prevalence of the d3-GHR isoforms was 50% but differed among the groups ($p = 0.006$), with a high prevalence (88%) in the group born SGA with verified intrauterine growth restriction. The d3-GRH allele was associated with decreased third-trimester FGV ($p = 0.05$) in SGA subjects. In the entire cohort, carriers of the d3-GHR allele had a significantly increased height (-0.10 vs. 0.34 SD score; $p = 0.017$) and change in height from birth to adolescence compared with carriers of the full-length GHR allele (0.57 vs. -0.02 SD score; $p = 0.005$). In conclusion, an increased spontaneous postnatal growth velocity was present in the carriers of the d3-GHR allele. Interestingly, the opposite effect on prenatal growth was detected in the SGA group, with a decreased FGV in carriers of the d3-GHR allele.

In another study, it was hypothesized that nocturnal GH concentrations, basal IGF-I and IGFBP-3 levels, and insulin sensitivity might show variations among individuals depending on their GHR allelic variant. To test this hypothesis, 38 prepubertal low birth weight children were studied with nocturnal GH concentrations, IGF-I and IGFBP-3 levels and insulin sensitivity during OGTT and insulin tolerance being measured. The GHR allelic variant was analyzed through multiplex PCR analysis in DNA from peripheral leukocytes. Characteristics of the overnight GH secretion [(mean GH: 6.8 ± 0.6 vs. 6.2 ± 0.5 ng/ml), (AUC: $3,227 \pm 280$ vs. $2,908 \pm 212$ ng/ml $\cdot$ min), (peak number: 4.4 ± 0.3 vs. 4.4 ± 0.2), (amplitude: 12 ± 1.1 vs. 10.8 ± 1.1 ng/ml)] did not differ between groups (f1/f1 vs. f1/d3 plus d3/d3). In addition, no significant differences were found in serum IGF-I SDS (-0.49 ± 0.26 vs. -0.40 ± 0.35) or

IGFBP-3 SDS (−1.21 ± 0.24 vs. −0.89 ± 0.21) or in insulin sensitivity (WIBSI: 12 ± 1.2 vs. 10.8 ± 1.1) in low birth weight children with full-length GHR compared to children carrying at least one GHRd3 allele. The distribution of the f1/f1 allelic variant and fi/d3 or d3/d3 was similar in the low birth weight children with or without catch-up growth. These results suggest that the GHR allelic variant does not play a significant role in the regulation of the GH-IGF-I/BP3 axis or in insulin sensitivity in prepubertal low birth weight children [8].

Growth Hormone-Insulin-Like Growth Factor System Plays a Pivotal Role in Fetal Development and Maturation and Differentiation

Fetal growth depends upon substrate and oxygen supply, vascularization of placental and fetal tissues, and endocrine modulation of cellular proliferation and tissue expansion. IGF-I has been shown to be a key stimulus of placental substrate uptake. IGF-I inhibits fetal placental catabolism and reduces placental lactate production. IGF-I deletions cause intrauterine failure to thrive. It is known that approx. 10% of infants with IUGR remain small. The causes of their growth deficits have remained unknown. Recently, monoallelic loss of chromosome 15q, mutations of the IGF-I receptor gene and loss of one copy of the IGF-I receptor gene again due to deletions of the distal long arm of chromosome 15 have been found in patients with intrauterine growth retardation and postnatal growth deficit. Binding of IGF-I to erythrocytes in short children with IUGR has been shown to be lower than in children with normal height. The number of IGF-I receptor copies on human fibroblasts seems to be predictive of their proliferative response to IGF-I. Hemizygosity for IGF-IR can cause primary IGF-I resistance despite normal or even elevated GH and/or IGF-I serum concentrations. At present, IGF-dependent growth in prenatal life seems to be largely independent of GH, except for a small effect just before birth, while IGF-dependent growth after birth and particularly so during puberty is strongly related to GH action. In conclusion, mutations and deletions of the IGF-I receptor gene lead to abnormalities in the function and/or number of IGF-I receptors. They seem to retard intrauterine and subsequent growth in humans. In the future, expression of such mutations in cells in vitro provides an opportunity to define the role of IGF-I receptor in human growth and growth disorders [9–12].

The IGF-IR is an important mediator of cell proliferation and longitudinal bone growth. The IGF-IR is closely related to the insulin receptor with a homology of 80–95% in the tyrosine kinase domain [9, 12]. IGF-I binding to the IGF-IR causes the transmembrane activation of the tyrosine kinase activity of the IGF-IR. In contrast to the insulin receptor, an impairment of the closely related IGF-IR has only been suggested on rare occasions under clinical circumstances: for instance, monoallelic loss of chromosome 15q, mutations of the IGF-I receptor gene and loss of one copy of the IGF-I receptor gene again due to deletions of the distal long arm of chromosome 15 have been found in patients with intrauterine growth retardation and postnatal growth deficit. Binding of IGF-I to erythrocytes in short children with IUGR has been shown to be lower than in children with normal height [9]. The number of IGF-I receptor copies on human fibroblasts seems to be predictive of their proliferative response to IGF-I. Hemizygosity for IGFIR can cause primary IGF-I resistance despite normal or even elevated GH and/or IGF-I serum concentrations. IGF-IR gene knockout experiments in mice have shown that mice carrying null mutations of both alleles exhibit severe and those of one allele moderate embryonic and postnatal growth deficiency [9, 13]. Recently, a compound heterozygous mutation of the human IGF-IR gene and a nonsense mutation (Arg59stop) that reduced the number of IGF-I receptors on fibroblasts from the affected child have been described in children with intrauterine growth retardation and highly elevated IGF-I levels [12]. The phenotype of patients with IGF-I receptor mutations and/or reduction of IGF-I receptor numbers known to date are being presented. Their relevance for clinical medicine and a hypothesized potential for future research on growth and growth disorders is being discussed.

Insulin-Like Growth Factors and Insulin Growth Factor-Binding Proteins

Disturbances in the GH/IGF-I axis are reported in 25–60% of short children born SGA [13]. It was hypothesized that these abnormalities might be related to abnormalities in the pituitary region. Therefore, magnetic resonance imaging (MRI) of short SGA children were compared to MRI results of other groups of short children and to normal controls in one recent study from Rotterdam. Pituitary height (PH) and thickness of the pituitary stalk (PS) were measured and their relationship with the maximum GH peak during a GH stimulation test, serum IGF-I and IGFBP-3 levels was evaluated. Short SGA children either with or without IGHD did not show major anatomical abnormalities in the hypothalamic-pituitary region in contrast to 58% of the non-SGA IGHD children and 87% of the short children with multiple pituitary deficiency (MPHD) who had anatomical abnormalities. PH in SGA children without GHD was normal whereas it was significantly lower in SGA children with IGHD. The lowest PHs were measured in non-SGA children with MPHD. A moderate decrease in PH was associated with significantly lower maximum serum GH peaks and lower serum IGF-I and IGFBP-3 levels. This study demonstrates that there is no indication to perform MRI of the pituitary region in short children born SGA without GHD [14].

Elevated fasting insulin levels and reduced insulin sensitivity in short SGA children was found to be related to elevated levels of overnight GH secretion in 16 short SGA children (age range 2.3–8.0 years) as compared to controls [15]. Compared with short normal-birth weight controls (n = 7, age range 2.3–5.0 years), short SGA children had higher fasting insulin levels (means: 26.8 vs. 20.6

pmol/l, p = 0.02), lower insulin sensitivity [means: 204 vs. 284% homeostasis model assessment (HOMA), p = 0.01], and higher β-cell function (112 vs. 89% HOMA, p = 0.04). SGA children also had lower levels of IGFBP-1 (87.0 vs. 133.8, p = 0.04), but similar IGF-I levels (IGF-I SDS: −1.1 vs. −1.7, p = 0.4). Compared with normal-height controls (n = 15, age range 5.6–12.1 years), SGA children had higher overnight GH secretion (GH maximum: 55.9 vs. 39.6 mU/l, p = 0.01; mean: 13.1 vs. 8.9, p = 0.004; minimum: 1.2 vs. 0.6, p = 0.02). The only hormone level significantly related to current height velocity was C-peptide (r = 0.75, p = 0.008). The authors hypothesized that resistance to the somatotropic actions of GH and IGF-I in short SGA children may contribute directly to reduced insulin sensitivity [15, 16].

Insulin-Like Growth Factors and Their Receptors and Binding Proteins

In the human, the IGF-I receptor gene is located on the distal long arm of chromosome 15 (15q26.3). The receptor is synthesized as a large precursor protein that undergoes extensive post-translational modifications including cleavage and glycosylation. The receptor belongs to the family of the IR and the IR-related receptor (IRR) [9]. The mature and functional IGF-I receptor is a heterotetramer, consisting of two alpha and two beta subunits. The alpha subunits form the extracellular domain for ligand binding. The IGF-I receptor binds IGF-I with high and IGF-II and insulin with lower affinity. In contrast, the insulin receptor is activated by low concentrations of insulin but higher doses of IGFs are required for activation of the IR. No ligand has been found for the IRR [9, 12]. The beta subunits of the receptors contain intracellular tyrosine kinase domains and are responsible for transphosphorylation of the receptors. Phosphorylation of the IGF-I receptor leads to interaction with a number of signaling molecules, phosphorylation of insulin-receptor substrates (IRS), activation of PI3-kinase (phosphatidyl-inositol-3) and MAP kinase [9, 12].

Growth during human fetal life represents the most rapid phase of human growth [10]. Fetal growth depends upon substrate and oxygen supply, vascularization of placental and fetal tissues, and a complex endocrine modulation of cellular proliferation, tissue expansion, inhibition of apoptosis and tissue remodeling. IGF-I and IGF-II have both been implicated to be important regulators of human fetal growth. IGF-I is produced by the fetal liver. This production is growth hormone independent but directly stimulated by insulin and fetal glucose uptake. IGF-I has been shown to be a key stimulus of placental substrate uptake. IGF-I inhibits fetal placental catabolism and reduces placental lactate production. Targeted disruption of the mouse IGF-II gene leads to a 40% reduction of fetal but normal postnatal growth. Disruption of the IGF-I gene leads to both pre- and postnatal growth failure. Mice with deletion of the IGF-I receptor gene have the most severe phenotype with birth weights of only 45%

Kiess · Kratzsch · Knüpfer · Robel-Tillig · Pulzer · Pfaeffle

of normal. Mice with an IGF-I receptor knock-out genotype usually die shortly after birth. Muscular hypotrophy leads to respiratory insufficiency in these animals pointing to the key role of the IGF-I receptor for the development and expansion of skeletal muscle [9–12].

IGF-I/GH Axis in Small-for-Gestational-Age Children: Impact on Growth and Metabolic and Cardiovascular Consequences

Epidemiological studies correlate low birth weight and the subsequent development of diabetes mellitus. A cohort study was carried out on 88 term infants (44 SGA and 44 AGA) and data on breastfeeding and age at weaning were registered at 1 year of age to identify the dietary and metabolic features associated with catch-up growth in infants born (SGA. In addition, anthropometric measurements, glucose, insulin, and leptin concentrations were measured at birth and at 1 year of age. A history of diabetes mellitus in a second-degree relative (p = 0.01) and complementary breastfeeding (p = 0.0003) were higher in SGA compared to AGA infants. Ten (13.6%) infants showed catch-up growth in length and weight combined. They had lower weight, glucose, insulin resistance index, and leptin concentrations at birth than those without catch-up growth. After logistic regression analysis for factors related to weight, gender, age at weaning, birth weight and leptin concentration at birth were included in the model (R^2 = 0.31; p = 0.00004). Female gender, early weaning, lower birth weight, and lower leptin concentration at birth were related to weight catch-up growth in these Mexican infants [17].

Low birth weight has been associated with an increased incidence of ischemic heart disease (IHD) and type 2 diabetes. Endocrine regulation of fetal growth by GH and IGF-I is complex. Placental GH is detectable in maternal serum from the 8th to the 12th gestational week, and rises gradually during pregnancy where it replaces pituitary GH in the maternal circulation. The rise in placental GH may explain the pregnancy-induced rise in maternal serum IGF-I levels. In the fetal compartment, IGF-I levels increase significantly in normally growing fetuses from 18 to 40 weeks of gestation, but IGF-I levels are four to five times lower than those in the maternal circulation. Thus, IGF-I levels in the fetal as well as in the maternal circulation are thought to regulate fetal growth. Circulating levels of IGF-I are thought to be genetically controlled and several IGF-I gene polymorphisms have been described [see chapter by Chernausek, this vol., pp. 44–59]. IGF-I gene polymorphisms are associated with birth weight in some studies but not in all. Likewise, IGF-I gene polymorphisms are associated with serum IGF-I in healthy adults in some studies, although some controversy exists. Serum IGF-I decreases with increasing age in healthy adults, and this decline could hypothetically be responsible for the increased risk of IHD with ageing. A recent nested case-control study found that adults without IHD, but with low circulating IGF-I levels and high IGF binding protein-3 levels, had a significantly increased risk of developing IHD during a 15-year follow-up period. In summary, the GH/IGF-I axis is involved in the regulation of fetal growth. Furthermore, it has been suggested that low IGF-I may increase the risk of IHD in otherwise healthy subjects. Hypothetically, intrauterine programming of the GH/IGF axis may influence postnatal growth, insulin resistance and consequently the risk of cardiovascular disease. Thus IGF-I may serve as a link between fetal growth and adult-onset disease [16–20].

It has been hypothesized that adiponectin levels as measured in neonatal dried blood spot samples (DBSS) might be affected by both prematurity and being born SGA. Therefore, adiponectin levels were measured in routinely collected neonatal DBSS taken on day 5 (range 3–12) postnatally from infants in a retrospective case-control study: in 122 infants (62 very premature (34 SGA) and 60 mature infants (27 SGA)) adiponectin concentrations were determined in stored neonatal DBSS using a sandwich immunoassay based on flow metric Luminex xMap technology. Adiponectin was measurable in all samples, and repeated measurements correlated significantly ($r = 0.94$). Adiponectin concentrations were negatively associated with both SGA ($B = -0.283$, $p = 0.04$) and prematurity ($B = -2.194$, $p < 0.001$), independently of each other. In the premature but not the mature group, adiponectin levels increased with increasing postnatal age at blood sampling ($B = 0.175$, $p < 0.001$). It was concluded that reliable quantification of adiponectin in stored DBSS is feasible and may be used to study large populations of routinely collected samples. Low levels of adiponectin in neonatal DBSS are associated with SGA as well as prematurity. Blood adiponectin levels increase with postnatal age in premature infants, suggesting a rapid yet unexplained metabolic adaptation to premature extrauterine life [21].

Being born SGA has been suggested to influence the female pituitary-gonadal axis, but only a few studies have focused on male pituitary-gonadal function. In one study, the objective was to evaluate the influence of fetal growth rate on male reproductive function. A follow-up study of a prospective study with data on third-trimester fetal growth velocity and birth weight was set up: 52 healthy adolescent males participated. They were divided into those born AGA and SGA, with or without intrauterine growth restriction. Pubertal stage, testicular size, and reproductive hormones were determined. Overnight 20-min LH and FSH profiles and overnight urine LH and FSH were determined in an additional study ($n = 30$). No significant differences were found in testosterone levels (19.2 vs. 18.9 nmol/l), inhibin B levels (186.5 vs. 188.0 pg/ml), or LH/testosterone ratio (0.15 vs. 0.18) between AGA and SGA, respectively. No significant differences in overnight secretory patterns of gonadotropins or testicular size and morphology were determined by ultrasonography between AGA and SGA. Fetal growth velocity did not influence any of the reproductive hormone levels. Overnight urinary LH and FSH excretion correlated statistically significantly with overnight LH ($r = 0.50$; $p = 0.02$) and FSH ($r = 0.44$; $p = 0.04$) secretion, respectively. In conclusion, poor third-trimester growth and/or low birth weight had no effect on subsequent male reproductive hormones. Contrasting a previous studies, no difference in testosterone or inhibin B levels between SGA and AGA were found, suggesting that testicular function was not impaired in adolescent males born after compromised fetal growth [7].

Growth Hormone Therapy

Recently, the International Societies of Pediatric Endocrinology and the Growth Hormone Research Society have published their views on the management of the child born small for gestational age and in particular the potential use of GH therapy in SGA children: low birth weight remains a major cause of morbidity and mortality in early infancy and childhood. It is associated with an increased risk of health problems later in life, particularly coronary heart disease and stroke. There was consensus that diagnosis of SGA should be based on accurate anthropometry at birth including weight, length, and head circumference. They recommended early surveillance in a growth clinic for those children without catch-up. Early neurodevelopment evaluation and interventions are warranted in at-risk children. Endocrine and metabolic disturbances in the SGA child are recognized but infrequent. For the 10% who lack catch-up, GH treatment can increase linear growth. Early intervention with GH for those with severe growth retardation (height SDS, <-2.5; age, 2–4 years) should be considered at a dose of 35–70 μg/kg $\times$ day. Long-term surveillance of treated patients is essential. The associations at a population level between low birth weight, including SGA, and coronary heart disease and stroke in later life are recognized, but there is inadequate evidence to recommend routine health surveillance of all adults born SGA outside of normal clinical practice [22].

Mean total IGF-I in short, SGA children is reduced, but within the normal range. Free/dissociable IGF-I is considered to be the bioactive form of IGF-I. Therefore, in one study changes in free IGF-I during GH treatment in short SGA children were measured. It was asked whether or not free IGF-I levels contributed to predicting first-year growth response and/or adult height in a randomized, double-blind GH dose-response study with a GH dose of either 1 mg/m^2 · day (group A) or 2 mg/m^2 · day (group B). Free IGF-I, total IGF-I, and IGFBP-3 were determined at baseline, after 1 and 5 years, at stop, and 6 months after GH discontinuation. 73 (46 male) short SGA children (36 group A) with a baseline mean age of 7.7 (2.2) years and a mean GH duration of 8.2 (2.1) years were studied. Untreated SGA children had a mean free IGF-I SDS of -0.2 (1.2), not related to total IGF-I. During GH therapy, free IGF-I significantly increased to 1.6 (0.7) SDS, as did total IGF-I and IGFBP-3 (2.0 (0.8) and 1.3 (0.9), respectively). Multiple regression analysis showed that baseline free IGF-I and IGFBP-3 were negatively correlated with adult height SDS, whereas baseline bone age delay, target height SDS, baseline height SDS, and GH dose were positively correlated. Free IGF-I was also negatively correlated with first-year growth response. The authors concluded that circulating baseline free IGF-I and IGFBP-3 were better predictors for adult height in GH-treated SGA children than total IGF-I, or total IGF-I to IGFBP-3 ratio. This suggests a possible role for free IGF-I measurement in predicting the effect of GH therapy in short SGA children [19].

Conclusion

Interventions that affect maternal and child undernutrition and nutrition-related outcomes include: promotion of breastfeeding; strategies to promote complementary feeding, with or without provision of food supplements; micronutrient interventions; general supportive strategies to improve family and community nutrition; and reduction of disease burden (promotion of handwashing and strategies to reduce the burden of malaria in pregnancy). Although strategies for breastfeeding promotion have a large effect on survival, their effect on stunting is small. In populations with sufficient food, education about complementary feeding increased height-for-age Z score by 0.25 (95% CI 0.01–0.49), whereas provision of food supplements (with or without education) in populations with insufficient food increased the height-for-age Z score by 0.41 (0.05–0.76) [23]. Management of severe acute malnutrition according to WHO guidelines reduced the case-fatality rate by 55% (risk ratio 0.45, 0.32–0.62), and recent studies suggest that newer commodities, such as ready-to-use therapeutic foods, can be used to manage severe acute malnutrition in community settings. Effective micronutrient interventions for pregnant women included supplementation with iron folate (which increased hemoglobin at term by 12 g/l, 2.93–21.07) and micronutrients (which reduced the risk of low birth weight at term by 16% (relative risk 0.84, 0.74–0.95). Recommended micronutrient interventions for children included strategies for supplementation of vitamin A (in the neonatal period and late infancy), preventive zinc supplements, iron supplements for children in areas where malaria is not endemic, and universal promotion of iodized salt. Existing interventions that were designed to improve nutrition and prevent related disease could reduce stunting at 36 months by 36%; mortality between birth and 36 months by about 25%; and disability-adjusted life-years associated with stunting, severe wasting, intrauterine growth restriction, and micronutrient deficiencies by about 25%. To eliminate intrauterine growth retardation and/or stunting in the longer term, these interventions should be supplemented by improvements in the underlying determinants of undernutrition, such as poverty, poor education, disease burden, and lack of women's empowerment [23]. The high mortality and disease burden from intrauterine growth retardation caused by nutrition-related factors make a strong case for urgent implementation of preventive interventions [24]. In many countries around the world, GH treatment of SGA children cannot be afforded by many. Simple preventive measures are more ethical and in addition more likely to be affordable by the social systems in many countries.

Kiess · Kratzsch · Knüpfer · Robel-Tillig · Pulzer · Pfaeffle

References

1 Bukowski R, Smith GC, Malone FD, Ball NH, Nyberg DA, Comstock CH, Hankins GD, Berkowitz RL, Gross SJ, Dugoff L, Craigo SD, Timor-Tritsch IE, Carr SR, Wolfe HM, D'Alton ME: FASTER Research Consortium Fetal growth in early pregnancy and risk of delivering low birth weight infant: prospective cohort study. BMJ 2007;334:836–842.

2 Eide MG, Oyen N, Skjaerven R, Bjerkedal T: Associations of birth size, gestational age, and adult size with intellectual performance: evidence from a cohort of Norwegian men. Pediatr Res 2007;62: 636–642.

3 Saigal S, Doyle LW: An overview of mortality and sequelae of preterm birth from infancy to adulthood. Lancet 2008;371:261–269.

4 Chernausek SD: Mendelian genetic causes of the short child born small for gestational age. J Endocrinol Invest. 2006;29(suppl 1):16–20.

5 Pagani S, Chaler EA, Radetti G, Travaglino P, Meazza C, Bozzola E, Sessa N, Belgorosky A, Bozzola M: Variations in biological and immunological activity of growth hormone during the neonatal period. Horm Res 2007;68:145–149.

6 Audi L, Esteban C, Carrascosa A, Espadero R, Perez-Arroyo A, Arjona R, Clemente M, Wollmann H, Fryklund L, Parodi LA, Spanish SGA Study Group: Exon 3-deleted/full-length GH receptor polymorphism genotype frequencies in Spanish small-for-gestational-age children and adolescents and in an adult population show increased fl/fl in short SGA. J Clin Endocrinol Metab 2006;91:5038–5043.

7 Jensen RB, Vielwerth S, Larsen T, Greisen G, Leffers H, Juul A: The presence of the d3-GH receptor polymorphism is negatively associated with fetal growth but positively associated with postnatal growth in healthy subjects. J Clin Endocrinol Metab 2007;2758–2763.

8 Mericq V, Roman R, Iniguez G, Angel B, Salazar T, Avila A, Perez-Bravo F, Cassorla F: Relationship between nocturnal GH concentrations, serum IGF-I/IGFBP-3 levels, insulin sensitivity and GH receptor allelic variant in small for gestational age children. Horm Res 2007;68:132–138.

9 Kiess W, Kratzsch J, Keller E, Schneider A, Raile K, Klammt J, Seidel B, Garten A, Schmidt H, Pfäffle R: Clinical examples of disturbed IGF signaling: intrauterine and postnatal growth retardation due to mutations of the insulin-like growth factor I receptor (IGF-IR) gene. Rev Endocr Metab Disord 2005;6:183–187.

10 Rosenfeld RG: Insulin-like growth factors and the basis of growth. N Engl J Med 2003;349:2184–2186.

11 Schmidt A, Chakravarty A, Brommer E, Fenne BD, Siebler T, De Meyts P, Kiess W: Growth failure in a child showing characteristics of Seckel syndrome: possible effects of IGF-I and endogenous IGFBP-3. Clin Endocrinol (Oxf) 2002;57:293–299.

12 Abuzzahab MJ, Schneider A, Goddard A, Grigorescu F, Lautier C, Keller E, Kiess W, Klammt J, Kratzsch J, Osgood D, Pfäffle R, Raile K, Seidel B, Smith RJ, Chernausek SD, Intrauterine Growth Retardation (IUGR) Study Group: IGF-I receptor mutations resulting in intrauterine and postnatal growth retardation. N Engl J Med 2003;349:2211–2222.

13 Saenger P, Czernichow P, Hughes I, Reiter EO: Small for gestational age: short stature and beyond. Endocrine Rev 2007;28:219–251.

14 Arends NJ, Van de Lip W, Robben SG, Hokken-Koelega AC: MRI findings of the pituitary gland in short children born small for gestational age (SGA) in comparison with GH-deficient (GHD) children and children with normal stature. Clin Endocrinol 2002;57:719–724.

15 Woods KA, van Helvoirt M, Ong KK, Mohn A, Levy J, de Zegher F, Dunger DB: The somatotropic axis in short children born small for gestational age: relation to insulin resistance. Pediatr Res 2002;51:76–80.

16 Jensen RB, Chellakooty M, Vielwerth S, Vaag A, Larsen T, Greisen G, Skakkebaek NE, Scheike T, Juul A: Intrauterine growth retardation and consequences for endocrine and cardiovascular diseases in adult life: does IGF-I play a role? Horm Res 2003;60(suppl 1):136–148.

17 Amador-Licona N, Martinez-Cordero C, Guizar-Mendoza JM, Malacara JM, Hernandez J, Alcala JF: Catch-up growth in infants born small for gestational age – a longitudinal study. J Pediatr Endocrinol Metab 2007;20:379–386.

18 Jensen RB, Vielwerth S, Larsen T, Greisen G, Veldhuis J, Juul A: Pituitary-gonadal function in adolescent males born appropriate or small for gestational age with or without intrauterine growth restriction. J Clin Endocrinol Metab 2007;92:1353–1357.

19 Bannink EM, van Doorn J, Mulder PG, Hokken-Koelega AC: Free/dissociable IGF-I, not total IGF-I correlates with growth response during GH treatment in children born small for gestational age. J Clin Endocrinol Metab 2007;92:2992–3000.

20 Brabant G, von zur Mühlen A, Wüster C, Ranke MB, Kratzsch J, Kiess W, Ketelslegers JM, Wilhelmsen L, Hulthén L, Saller B, Mattsson A, Wilde J, Schemer R, Kann P, German KIMS Board: Serum insulin-like growth factor I reference values for an automated chemiluminescence immunoassay system: results from a multicenter study. Horm Res 2003;60:53–60.

21 Klamer A, Skogstrand K, Hougaard DM, Nørgaard-Petersen B, Juul A, Greisen G: Adiponectin levels measured in dried blood spot samples from neonates born small and appropriate for gestational age. Eur J Endocrinol 2007;157:189–194.

22 Clayton PE, Cianfarani S, Czernichow P, Johannsson G, Rapaport R, Rogol A: Management of the child born small for gestational age through to adulthood: a consensus statement of the International Societies of Pediatric Endocrinology and the Growth Hormone Research Society. J Clin Endocrinol Metab 2007;92:804–810.

23 Bhutta ZA, Ahmed T, Black RE, Cousens S, Dewey K, Giugliani E, Haider BA, Kirkwood B, Morris SS, Sachdev HP, Shekar M, Maternal and Child Undernutrition Study Group: What works? Interventions for maternal and child undernutrition and survival. Lancet 2008;371:417–440.

24 Black RE, Allen LH, Bhutta ZA, Caulfield LE, de Onis M, Ezzati M, Mathers C, Rivera J, Maternal and Child Undernutrition Study Group: Maternal and child undernutrition: global and regional exposures and health consequences. Lancet 2008;371:243–260.

Prof. Wieland Kiess, MD
Hospital for Children and Adolescents, University of Leipzig
Liebigstrasse 20a
DE–04103 Leipzig (Germany)
Tel. +49 341 97 26 000, Fax +49 341 97 26 009, E-Mail wieland.kiess@medizin.uni-leipzig.de

Kiess · Kratzsch · Knüpfer · Robel-Tillig · Pulzer · Pfaeffle

Kiess W, Chernausek SD, Hokken-Koelega ACS (eds): Small for Gestational Age. Causes and Consequences.
Pediatr Adolesc Med. Basel, Karger, 2009, vol 13, pp 99–115

Clinical Management of Small-for-Gestational-Age Babies

Matthias Knüpfer

Department of Neonatology, Hospital for Children and Adolescents,
University of Leipzig, Leipzig, Germany

Abstract

As is well known, the long-term outcome of small-for-gestational-age (SGA) children is impaired, when compared to appropriate-for-gestational-age (AGA) children. Opposite to the long-term outcome, this article deals with the immediate postnatal clinical situation of preterm SGA children and shows that these children need some special considerations. The article tries to give rules for the termination of gestation and the preferable way of delivery. Immediately after delivery many data show that SGA preterm infants have more complications compared to AGA children. When analyzing development in the first postnatal days, it has to be stated that SGA children generally have more difficulties compared to their AGA counterparts. First, this article analyses the literature to get an analysis of the particularities of SGA children. We then describe our own experiences with the management of this patient group. Some clinical recommendations for diagnostic and therapeutic strategies are given at the end of each section.   Copyright © 2009 S. Karger AG, Basel

We here describe the postnatal course of small-for-gestational-age (SGA) children. Firstly, we are interested in the analysis of the clinical situation and try to describe the difficulties and the dangerous situations. After a discussion of the literature, we describe our own experiences in the management of SGA newborns. At the end of the each section, we will give some advice for the clinical management including treatment recommendations.

In this paper, we focus on the immediate postnatal development of SGA children in industrialized countries like Germany. Care of these children in developing countries is completely different. Special features of the care provided in developing countries are examined in another chapter of this book.

Postnatal Clinical Management of SGA Newborns: The Preterm Infants in Focus

The long-term outcome of SGA newborns, especially that of SGA preterm newborns, is worse compared to appropriate-for-gestational-age (AGA) newborns [1–3]. Apart from these long-term problems, the early postnatal care of SGA infants is also more difficult compared to AGA [1, 2, 4–8].

According to the Barker [9] hypothesis, mature SGA newborns may have a lot of long-term problems, but in the first days of life they are often not ill. On the contrary, SGA children born preterm have a lot of problems. There are two main reasons for this: Firstly, compared to SGA children who reach maturity, it is likely that intrauterine chronic malnutrition is much higher in preterm SGA children. This leads to higher intrauterine growth retardation (IUGR) and may trigger preterm delivery. Secondly, the preterm children additionally suffer from the immaturity itself. Therefore, preterm SGA newborns are stressed in two ways, first by the fact that they are SGA children and second by the immaturity. Since term SGA children mostly do not have severe problems immediately after birth, the article focuses on preterm SGA children and describes our management of this special patient group.

Who Is SGA and Why?

The central question for the detection of SGA patients has been answered differently: Some authors use a growth below the 10th percentile [1, 4], some below the 5th [5], and some use the 3rd [10] or the 1st percentiles [2]. Therefore, the data of these authors should be handled with care, since the authors sometimes do not investigate the same patient populations. For our clinical work, we consider a newborn as SGA when the birth weight is below the 10th percentile. In recent decades, a couple of authors [6, 11] worked on the SGA problem and described the differences between the problem of IUGR and low birth weight. The reasons for the latter are widespread and include chromosomal aberrations, virus infections and others (table 1), but these causes of low birth weight are rare. Most common and therefore most important is the growth restriction due to a reduced utero-placental-fetal blood flow, followed by a decreased nutrition of the fetus. As a consequence, a lot of authors included only these children and excluded other fetuses which were born as SGA due to other reasons. In this paper, we also exclude the latter and focus on children with intrauterine malnutrition.

We regularly observe different pathophysiological changes in these fetuses, which are described in table 2. There is a widespread pathology ranging from a fetus with a minimal growth restriction and no other problems at birth, a fetus with a high grade of intrauterine growth restriction with different organ pathology, to a stillbirth case. These different stages have one uniform reason: the insufficiency of the maternal-fetal axis. Using ultrasound, an experienced obstetrician is able to measure the blood flow profile in maternal, placental and fetal vessels to describe the exact state of the fetus in only one session. Multiple investigations describe the dynamics of the fetal development and help to answer the most important question: When to deliver?

Table 1. Factors leading to IUGR

Maternal factors	maternal diseases like cardiac or pulmonary dysfunction
	pregnancy diseases: hypertonus, HELLP
	abuse/intoxication: alcohol, nicotine, drugs
	demographic/socioeconomic/constitutional factors: multiple gestations, low socioeconomic status, mother's extreme age, parental height
Uterine/placental factors	placental infarcts
	placental insufficiency
	uterine abnormalities
Fetal factors	chromosomal abnormalities and gene defects
	infections

Table 2. Chronic intrauterine malnutrition

Chronic insufficiency of the utero-placental-fetal axis leads to
–Low fetal blood flow with chronic reduced nutrition status (oxygen, glucose, protein, fatty acid metabolism)
–Initiation of the so-called 'brain-sparing mechanism' to maintain the blood flow and nutrition of brain, coronary arteries, adrenal gland

As a consequence

Other fetal organs have a lower blood flow
–This leads to malnutrition
–This could have an impact on the postnatal course and especially on the intestine, the kidney and the liver, spleen and bone marrow
–The fetus increases the erythropoiesis to compensate the low oxygen status
–There is increased fetal stress situation with upregulation of glucocorticoids

Delivery – Where, How and When?

Once intrauterine growth retardation has been diagnosed, mother and fetus should be monitored closely by fetal biometry and Doppler investigations of uterine and fetal vessels. The risk of perinatal mortality of SGA infants is up to 20 times higher compared to AGA infants [12]. It has been shown that SGA children have a higher rate of low Apgar scores (5-min Apgar <3), umbilical artery acidosis (pH <7.0) and a higher rate for intubation at delivery [10].

The first question – Where to deliver – is most easily answered: because of the increased risk of perinatal complications, SGA fetuses with 34.0 weeks of gestational age or below and fetuses with a high grade of IUGR independent of gestational age should be delivered in tertiary centers to provide immediate admission to a neonatal intensive care unit to manage short-term and long-term complications in these children [8].

The second question – How to deliver? – is more difficult to answer. Close cooperation between obstetricians and neonatologists is essential to decide how and when to deliver. It is sometimes difficult to decide which mode of delivery is the better one: in vaginal delivery, uterine contractions may increase the hypoxic stress further, and a primary cesarean section may lead to a disturbed cardiopulmonary adaption, i.e. wet lung syndrome. It might be interesting that some studies with preterm SGA have shown that in general SGA preterm newborns have a better lung maturity than AGA preterm infants [for details, see chapter by Chernausek, this vol., pp. 44–59]. In view of the latter, the decision for a primary cesarean section might be easier. However, most of the SGA fetuses do not need to be delivered by a cesarean section. They can be delivered vaginally.

The third question – When to deliver? – seems to be the most difficult. The answer should include the aspect of additional high risk of the premature infant with immature organs, which might worsen the starting conditions on the one hand [8] and the dangerous intrauterine stay of the fetus with a high risk malnutrition status on the other hand. The latter includes the risk of a stillbirth, which is notably increased in an IUGR fetus [2, 8]. A comprehensive European study shows a wide range of disagreement about the correct time of delivery, and a willingness to randomize patients to clinical trials of management [13], but until today no randomized data are available.

Thus, it remains very difficult to give general rules for the delivery time point. Of course, the fetus should be delivered in case of fetal decompensation with pathological cerebral perfusion. A short time ultrasound check should take place in case of fetal compensation with pathological perfusion of the fetal umbilical artery. A small German study clearly shows that a fetus with impaired blood flow in the umbilical artery has a notably higher short- and long-term mortality and morbidity compared to controls [14].

Some analyses show a close correlation between morbidity/mortality and gestational age of SGA children. The risk of dying or developing a severe adaptation disturbance is very high before the 28th gestational week, afterwards the risk decreases, but after the 34th week of gestation, the risk rises again. The explanation seems easy: in the first group the IUGR is combined with a high grade of immaturity, and in the latter group it may be that a fetus with severe IUGR stays in the uterus too long.

Especially in low-gestational-age children (<30 weeks), the problem of severe growth restriction and severe immaturity remains. Some elucidation can be found in the data of Aucott et al. [5]. They compared SGA infants of nearly 30 weeks gestational age with AGA preterm infants having the same birth weight but a lower maturity (25.5 weeks). The mortality was higher in the low-gestational AGA children (30 vs. 20.5%) and the surviving SGA children born at nearly the 30th gestational week had less intraventricular hemorrhage (IVH) and periventricular leukomalacia (PVL) than their immature counterparts, but a similar incidence of retinopathy of prematurity

(ROP). However, the SGA preterm infants had a higher incidence of necrotizing enterocolitis (NEC), direct hyperbilirubinemia and chronic lung disease. Therefore, when considering an elective preterm delivery for these high-risk children, the increased risks in the neonatal period should be taken into account. Below 28 weeks of gestational age, the fetus should be delivered only in case of emergency. After 32–34 weeks, the intrauterine stay should be prolonged only if the fetus really profits from it. This mostly means ongoing growth over time measured in repeated ultrasound investigations.

For observation of the fetus, the mother's help should not be underestimated: decreased or absent fetal movements are a strong sign of fetal emergency. If the mother describes decreased fetal movements, we strongly advise further investigation using fetal cardiotocography (CTG) or/and a fetal ultrasound.

To make a decision, we need an experienced obstetrician with special knowledge in fetal, maternal and placental ultrasound diagnostics, a current CTG, and of course an attentive mother for observation of the fetal movements.

Clinical Recommendations

To conclude, we advise that a preterm SGA <34th weeks' GA should be delivered at a tertiary center; mostly, these children can be delivered vaginally, however the indication for a primary cesarean section is wide. A delivery below 28–30 gestational weeks should be handled with care and should only be initiated if there is a fetal (or maternal) emergency. If the 34th gestational week has been reached, the fetus should be kept in utero only if it really profits from this stay. The delivery should be as atraumatic as possible, since additional stress for the SGA stressed fetus may lead to fetal decompensation.

Adaptation of the Lung – Risk of Respiratory Distress Syndrome and Bronchopulmonary Dysplasia

One of the most dangerous adaptation problems of preterm infants is respiratory distress syndrome (RDS) and long-term lung damage with the development of bronchopulmonary dysplasia (BPD), or chronic lung disease (CLD). Some studies compare the RDS rate in SGA and AGA preterm infants. The results are controversial. Some studies show a higher RDS rate in SGA infants [15, 16]. It has been speculated that SGA children have an increased vulnerability of the extreme immature lung [10]. Interestingly, other authors found a decreased rate of RDS in SGA preterm infants. These authors discuss the accelerated fetal pulmonary maturation because of the permanent intrauterine stress situation [1, 4]. It may be that the analysis at different time points produces different results: when analyzing the maturity (<28 weeks GA) of low-gestational-age SGA/AGA pairs, acceleration of pulmonary maturation plays a much more important role and leads to a better postnatal status of the SGA, whereas in higher-gestational-age preterm infants this fact seems to have a lower impact. Besides these different data about RDS development, the long-term outcome studies are unique: SGA preterm children have a worse outcome. In a couple of studies, the BPD rate was around 2-fold higher in SGA compared to AGA preterm infants [15, 17, 18].

Clinical Recommendations

The clinical management of respiratory therapy is not different between SGA and AGA children. An early CPAP application is advised immediately after birth in case of RDS. In our NICU, surfactant is not given in the delivery room. First the child is treated to get a stable circulation status. If necessary we support with volume and catecholamines. Before surfactant application, a left-to-right shunt of the patent ductus arteriosus (PDA) should be documented using echocardiography. Using CPAP (up to 6–7–8 cm H_2O), we try to prevent mechanical ventilation. It is extremely important to make gentle ventilation if mechanical ventilation is needed. The use of diuretics (furosemide, thiazide and spironolactone) is discussed controversially [19, 20]. We give furosemide (0.5–1 mg/kg/day) and spironolactone (1–2 mg/kg/day) in selected cases. Postnatal corticosteroids may be effective but, due to their severe side effects, should be restricted to the severe cases using a single dose of 0.2–0.3 mg of dexamethasone.

Adaptation of the Heart and Circulation

A stable circulation is necessary for all children, but extremely important in SGA newborns, because these children need a sufficient perfusion in the formerly less perfundated organs for the restitution of organ function and to wash out metabolic products, especially lactate.

In comparison with this very important field, little is known about the heart and circulation status in SGA preterm infants. Available data [21] show that SGA preterm infants have a reduced myocardial contractility compared to their AGA counterparts, which leads to a lower stroke volume. On the other hand, these children have a significantly higher heart rate and if taken together they reach a higher cardiac output per minute (stroke volume × heart rate) [21].

A second main aspect in the circulation is the higher incidence of a patent ductus arteriosus with a significant left-to-right shunt in SGA newborns [5, 21, 22]. This will additionally worsen the clinical situation because the left-to-right shunt shifts blood into the lung. This may lead to a lower perfusion of the organs in the main circulation. Brain perfusion is crucial for the child's prognosis and pathological flow pictures (as shown in figure 1) should be strictly avoided to defend the cerebral integrity. Furthermore, a PDA may contribute to a prolonged lower perfusion of the gastrointestinal organs, the kidney and the skin.

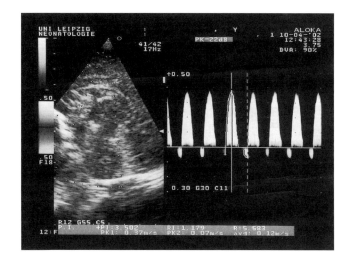

Fig. 1. Pathologic cerebral perfusion with a reverse flow in the anterior cerebral artery.

Clinical Recommendations

In preterm children, an early echocardiography (within 6 h after birth) is performed to assess the circulatory status and myocardial contractility. If necessary, a support by dobutamine (6–8–10 μg/kg/min) is recommended. An appropriate cardiac output (above 200 ml/kg/min) should show a normal perfusion index at least in the anterior cerebral artery. Hypovolemia is treated using isotonic electrolyte solutions (fractionated, 10–20 ml/kg). Especially in low gestational age children (<28 weeks gestational age) this volume substitution should be given slowly over 30 min or 1 h to prevent excessive increased blood pressure transmitted to fragile cerebral capillaries, leading to intracranial hemorrhage. The circulatory status should be monitored closely by continuously measuring the central-peripheral temperature difference (central, measured on the back and peripheral, measured plantar). A difference above 2–2.5°C indicates significant centralization and needs further investigation of the circulatory status.

From the first postnatal day on, the flow of the PDA should be monitored closely. A hemodynamically significant PDA is treated with indomethacin 0.2 mg/day twice a day or ibuprofen 10 mg/kg. A hemodynamic significant PDA is defined as a reduced perfusion of the brain (anterior cerebral artery), the kidney (renal artery) or the intestine (superior mesenteric artery) due to the PDA steel phenomenon on the one hand and/or a cardiac, especially left ventricle hemodynamic stress situation on the other.

Metabolism: Glucose and Lactate

The fetal brain-sparing circulation leads to a reduced perfusion of other organs. This might be of great importance for the intestine, the kidney, the liver, spleen and bone marrow and the skin. Because of this reduced organ perfusion, SGA children mostly develop a chronic lactate acidosis. The lactate acidosis may rise in the first postnatal hours [23]. The reason for this is simple: the formerly low perfundated organs are reperfundated, and therefore the local metabolic acidosis is washed out and the general acidosis may be increased.

The most important postnatal metabolic problem is the hypoglycemia. A hypoglycemia <2.6 mmol/l is considered as relevant and dangerous according to the data of Lucas et al. [24]. It has been shown that even term SGA children had significantly higher rates of symptomatic hypoglycemia [25]. One may speculate that low glycogen storage is the main reason for the low blood glucose, but low body fat storage seems to be at least equally important [23].

Clinical Recommendations

An early intravenous line and the application of glucose of around 5 mg/kg/min is advised. It is necessary to observe blood glucose levels carefully to prevent hypoglycemia. In case of hypoglycemia, glucose 0.3–0.5 mg/kg should be substituted as a bolus. The daily glucose substitution should be increased correspondingly. If lactate acidosis rises in the first few hours after birth, we do not give bicarbonate first. Mostly, acidosis will disappear with an adequate circulation therapy. Bicarbonate therapy might be useful if lactate acidosis secondarily increases after an initial drop. Hyperglycemia with glucose levels above 10 μmol/l should be treated with insulin. The total daily glucose input should not fall below 5 g/kg/day.

Blood and Host Defense: Polyglobulia, Thrombocytopenia, Leukocytopenia, Sepsis Rate

The SGA preterm infants often exhibit polyglobulia, thrombocytopenia and leukocytopenia [26, 27]. There are two main possible reasons for the thrombocytopenia and leukocytopenia. First, the blood-forming organs belong to the intrauterine lower perfundated organs and therefore have a low capacity to supply the organism with leukocytes and thrombocytes. Second, as a compensation to the intrauterine malnutrition most of the SGA children have an increased hemoglobin concentration and erythrocyte count, resulting in a higher hematocrit level. As a consequence, this overproduction may lead to a suppressed production of thrombocytes and leukocytes. As expected, high levels of erythropoietin are found in cord blood of SGA newborns, mediating the high hematocrit level. In contrast, the thrombopoietin levels are normal. The latter indicates that the low thrombocyte count is not mediated by a low thrombopoietin level [27]. Also, animal trials have shown that chronic hypoxia upregulates erythropoiesis and downregulates leukopoiesis and thrombopoiesis, but the mechanisms involved in this regulation have not fully been elucidated [28].

The suppression of leukopoiesis and thrombopoiesis may persist over some days or even weeks postnatally. According to the described pathology, a bone marrow biopsy is not necessary in almost all the cases.

In case of thrombocytopenia, thrombocyte substitution is possible. For leukocytopenia, some data show that colony stimulation factors (G-CSF or GM-CSF) might be helpful [29]. According to the low leukocytes levels (and other factors), the SGA children might have a higher sepsis rate. Indeed, some authors found an increased sepsis rate, but the available data are not unique. Simchen et al. [30] found a higher sepsis rate, whereas the difference in a great American study [10] was only marginal (blood culture proven sepsis 1.7 vs. 1.2% in SGA and AGA, respectively).

Clinical Recommendations

In SGA children, polyglobulia is expected and therefore placental transfusion should be viewed critically in this patient group. A polyglobulia with venous hematocrit >0.7 should be treated by an isovolumetric exchange transfusion. If thrombocytopenia falls below 20 Gpt/l thrombocytes (20 ml/kg) should be substituted. Thrombocyte counts between 50 and 20 should be controlled every (4)-6–8–(12) h. For leucopenia, it is theoretically possible to give colony stimulation factors, but until today we have no experience with these agents, e.g. GCSF or GMCSF. Some data show that this might be useful, especially in preventing neonatal sepsis [29].

Brain: Risk of Intraventricular Hemorrhage and Periventricular Leukomalacia

In view of the neurology of SGA children compared to AGA newborns, different data have been published. An American study [10] found a 2-fold increase in severe IVH (IVH 3/4°) for preterm infants below the 3rd percentile (3.2 vs. 1.7% in AGA between 25th and 75th percentile), but the difference decreased when analyzing children below the 11th (2.4%) and above the 10th percentiles (1.7%). A comprehensive German study [4] found no differences for severe IVH, analyzing children between the 25th and 29th gestational weeks. In contrast, two other studies reported an increased risk of IVH in those SGA children born between 25 and 30 [31] or below 34 gestational weeks [7]. It might also be important to observe the frequency of PVL (fig. 2). A PVL is mostly due to a shortly absent or a severe reduced cerebral perfusion and this is typical for SGA children. There is a small, but not significant increase (1.5-fold) of the PVL cases in SGA [2], but no differences between SGA and AGA preterm infants in other studies [32, 33].

Besides these different research data, it is generally accepted that for all (but especially for preterm) children cerebral integrity is extremely important. Especially in extreme low birth weight preterm children, it is necessary to maintain normal circulation with a blood pressure in the normal range, since the brain perfusion is not autoregulated in these children. An abnormal pathological perfusion of the brain (fig. 1) is very dangerous in view of the development of IVH and PVL.

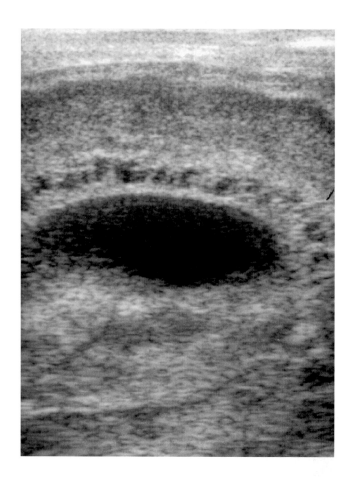

Fig. 2. Classical periventricular leukomalacia, which dramatically worsens the prognosis of the patient.

Clinical Recommendations

We strongly suggest choosing the most atraumatic delivery to protect the cerebral integrity. In the postnatal period, it is important to keep the circulation within the normal range and be aware of hypotonia and hypertonia. The aim is to guarantee a normal cerebral perfusion at any time. If ventilation is necessary (the ventilation should be a gentle ventilation) be aware of CO_2 levels below 30 mmHg. If pH is within the normal range (7.25–7.40), a slightly enhanced CO_2 (to 60–65 mmHg) may be tolerable (the so-called 'permissive hypercarnia'). However, some newer data show that the incidence of severe IVH increased with higher carbon dioxide values [34, 35]. Therefore, the idea of 'permissive hypercapnia' should be handled with care, today.

Adaptation of the Intestine: Nutrition and the Incidence of Gut Perforations and Necrotizing Enterocolitis

Figure 3 shows some of the conditions specific for SGA children, and should demonstrate that the development of gastrointestinal disturbances is a crucial point in the

The intestine itself has low gut motility, persistent abnormal blood perfusion, delay of meconium passage and viscous meconium	**Kidney** Persistent abnormal kidney perfusion means low urine production and edema

The arising disturbance of the intestine may lead to perforation and NEC

Heart and circulation Myocardial dysfunction and the high incidence of PDA result in reduced bowel perfusion	**Blood** High erythrocyte count results in polyglobulia with worse blood flow Low leukocyte count means low defending capacity

Fig. 3. The different pathophysiological pathways leading to deterioration of gastrointestinal disturbances.

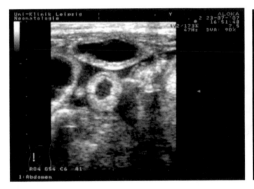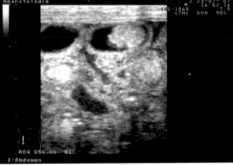

Fig. 4. Abdomen ultrasound of a child suffering from gastrointestinal motility disturbance. The SGA preterm newborn (27 weeks of gestation, 634 g) clinically showed abdominal distension and bilious vomiting. Using ultrasound very low gut motility with thick gut walls and free intra-abdominal liquid was found. The child developed a gut perforation and needed a surgical intervention on the 4th day of life.

pathology of SGA newborns. So it is no surprise that one of the most common problems in severe intrauterine growth-restricted children is the development of a postnatal gastrointestinal motility disturbance. Clinical signs are an abdominal distension, low gut motility, bilious vomiting, late meconium passage or even an absent meconium passage (fig. 4, 5). A lot of SGA children have a delay in tolerating enteral feeding

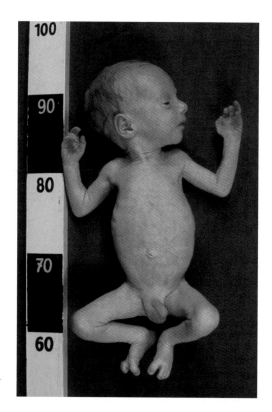

Fig. 5. SGA preterm child. Note the abdominal distension.

during the first days. It has been shown that these gastrointestinal adaptation problems are a direct result of the prenatal hemodynamic disturbances [36, 37]. However, some other pathological changes which were found in SGA children (hematological, host defense, circulation, PDA) could worsen the intestinal situation (fig. 6).

There are four possibilities after severe IUGR. First, a normal postnatal gastrointestinal adaptation can be observed. Second, the preterm children may have only a low adaptation disturbance with delayed build up of the enteral feeding. Third, these newborns may have a severe disturbance with long-term feeding intolerance but without the need for surgical intervention, and, fourth, the preterm infants may have a gut perforation or in the most severe case a NEC. In SGA preterm infants, the incidence of NEC is higher compared to AGA children (OR 2.47, CI 1.2–5.1).

It has been shown in a meta-analysis that feeding with donor human milk was associated with a significantly reduced relative risk of NEC [38]. If human milk is not successful, it is possible to try a hydrolyzed formula, e.g. Pregomin AS. A gut perforation/NEC is an emergency situation which mostly needs surgical intervention. In very immature preterm infants (<26th gestational weeks), we sometimes choose the conservative method: wait and antibiotics. The antibiotics should include a strong Gram-negative activity and an activity against anaerobic bacteria. This might be a

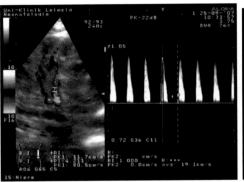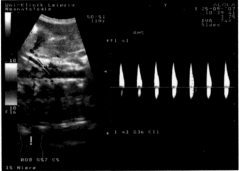

Fig. 6. Disturbed blood flow in two abdominal arteries: left renal artery (left) and superior mesenterial artery (right). These pictures are often seen in SGA children. The low perfusion is combined with an impaired function of the kidney and the gut. These disturbances may lead to a prolonged oliguria with elevated BUN and creatinine levels and a gastrointestinal motility disturbance with a high risk of the development of necrotizing enterocolitis.

carbapeneme or a combination between third-class cephalosporin (ceftazidime) and metronidazole.

It has been shown that the ultrasound investigation of blood flow parameters in the superior mesenteric artery (fig. 6, right) can be used to predict intestinal activity/motility [39] and early tolerance to enteral feeding [40]. Therefore, we suggest making an early investigation on the first day of life. A pulsatility index (PI) between 1.0 and 2.0 is considered normal, indicating a normal blood flow. A persistent hypoperfusion after birth (PI >2), or the opposite, a hyperperfusion (PI <1), are considered as abnormal perfusions. In these cases, we are very careful in the build up of feeding. Other authors advise the measurement of superior mesenteric artery flow velocity 30 min after the first feed as a good tool for predicting enteral feeding [40].

Clinical Recommendations

All preterm children with prenatal hemodynamic disturbances are carefully observed concerning abdominal distension and meconium passage during the first postnatal days.

We regularly measure the blood flow parameters in the superior mesenteric artery and use these results for evaluation of the risk of getting a severe gastrointestinal disturbance.

Oral feeding starts with tea 2–3 ml/kg every 3 h around 12 h after birth. Meconium passage might be induced using an external abdominal massage and a careful rectal infusion of 10–20 ml warm isotonic electrolyte solution. If the gut motility is low (clinical investigation), it can be increased with 0.1 mg/kg pyridostigmine intramuscularly. If meconium is very viscous one can add acetylcysteine 50 mg/kg into the rectal infusion solution. If meconium has been passed and the child tolerates tea, it might be useful to give a mixture of tea and human milk (1:1) for approximately 24 h. Afterwards, feeding is changed to pure raw human milk from the mother or a donor. We also have some experience with a hydrolyzed formula (Pregomin AS). Some of our data show that in children who do not tolerate human milk, the introduction of Pregomin AS seems to be an alternative. In case of vomiting, low-dose erythromycin (4–5 mg/kg/day i.v.) may increase the stomach propulsion. Antibiotics should be absolutely restricted in SGA children because of destruction of the normal intestinal flora. Daily monitoring of the intake of glucose, protein, fat and energy is essential.

Surgical intervention is mostly needed in cases of proved perforation or NEC. In very preterm children (<26 weeks' gestational age), the successful treatment with stop of nutrition and start of antibiotics is possible. If no surgical intervention takes place, the used antibiotic should include a strong activity against Gram-negative and anaerobic bacteria.

Adaptation of the Kidney

In IUGR infants, fetal growth of the kidneys is more impaired than the body as a whole [41]. This leads to a reduced kidney length at birth [42]. Aminoglycosides and glycopeptides are almost exclusively eliminated by renal excretion. These drugs may help to investigate the kidney function. Allegeart et al. [43] showed that renal drug clearance is significantly lower in preterm neonates born SGA than in AGA controls. This reduced clearance was observed not only at birth but also up to the postnatal age of 4 weeks. However, it has also been shown that kidney function in preterm SGA infants is not different compared to AGA preterm infants. In the same study, preterm SGA infants who received aminoglycoside therapy exhibit impaired glomerular and tubular function at 2 months of life [44].

Clinical Recommendations
Published results about the postnatal course of kidney function are rare, but available data show kidneys of SGA to be smaller compared to those of AGA children. Kidney function is decreased in SGA children. However, these differences do not really lead to a relevant decreased kidney function. As discussed above, the data of some groups show that the use of aminoglycosides should be handled with care in SGA children.

Conclusion and Treatment Recommendations

The postnatal situation of the SGA preterm children needs some special consideration. As a conclusion we advise:

(1) Make a good prenatal diagnostic, speak intensively with the obstetrician. Carefully use the obstetrician's information (including ultrasound and CTG). Do not forget the mother's information about the movement of the fetus.

(2) To make the best decision for the termination of the delivery, it is necessary to discuss the question: Will the fetus further profit from the intrauterine stay?

(3) Together with the obstetrician, choose the best kind of delivery. The stressed fetus does not need further stress during delivery.

(4) Insert an intravenous line for supporting the preterm with glucose and electrolytes. Carefully monitor the glucose serum level.

(5) The possible increase of lactate acidosis during the first postnatal day is due to the reperfusion of the formerly less perfundated organs and mostly does not need bicarbonate therapy. A combination of lactate acidosis and hypoglycemia in SGA newborns is mostly not a connatal disturbance of the metabolism. It is a result of the special situation of SGA newborns.

(6) Severe thrombocytosis and leukocytosis are sometimes present in SGA newborns. Mostly, this is not an infection, but of course it is necessary to look for CRP, IL-6 (or other) parameters to exclude an infection.

(7) We advise making an early ultrasound of the heart and a close clinical observation of the circulation. Use dobutamine and volume, if necessary. Measure the perfusion of the brain and try to maintain a normal cerebral perfusion.

(8) The gut is the most difficult problem of SGA preterm infants. We suggest being really careful with nutrition. It is necessary to carefully observe the status of the intestines. Try to estimate the perfusion status measuring the superior mesenteric artery. Start with tea intake at the end of the first postnatal day. Wait till meconium is seen. If meconium is not seen, be extremely carefully with start of nutrition. We try to induce the passage of meconium with careful rectal infusions. Investigate the child at least two times a day with special consideration being given to the abdomen.

References

1 Sharma P, Mc Kay K, Rosenkrantz TS, Hussain N: Comparison of mortality and pre-discharge respiratory outcomes in small-for-gestational-age and appropriate-for-gestational-age premature infants. BMC Pediatr 2004;4:1–7.

2 Kamoji VM, Dorling JS, Manktelow BN, Draper ES, Field DJ: Extremely growth-retarded infants: is there a viability centile? Pediatrics 2006;118:758–763.

3 Beinder E: Fetalzeit und spätere Gesundheit. Dtsch Arztebl 2007;104:A644–A650.

4 Bartels DB, Kreienbock L, Dammann O, Wenzlaff P, Poets CF: Population based study on the outcome of small for gestational age newborns. Arch Dis Child Fetal Neonatal Ed 2005;90:F53–F59.

5 Aucott SW, Donohue PK, Northington FJ: Increased morbidity in severe early intrauterine growth restriction. J Perinatol 2004;24:435–440.

6 Garite TJ, Clark R, Thorp JA: Intrauterine growth restriction increases morbidity and mortality among premature neonates. Am J Obstet Gynecol 2004;191:481–487.

7 Zaw W, Gagnon R, da Silva O: The risk of adverse neonatal outcome among preterm small for gestational age infants according to neonatal versus fetal growth standards. Pediatrics 2003;111:1273–1277.

8 Meyberg R, Boos R, Babajan A, Ertan AK, Schmidt W: Intrauterine growth retardation: perinatal mortality and postnatal morbidity in a perinatal center. Z Geburtshilfe Neonatol 2000;204:218–223.

9 Barker DJ: In utero programming of chronic disease. Clin Sci 1998;95:112–128.

10 McIntire DD, Bloom SL, Casey BM, Leveno KJ: Birth weight in relation to morbidity and mortality among newborn infants. N Engl J Med 1999;340:1234–1238.

11 Fang S: Management of preterm infants with intrauterine growth restriction. Early Hum Dev 2005;81:889–900.

12 Clausson B, Cnattingius S, Axelsson O: Preterm and term births of small for gestational age infants: a population-base study of risk factors among nulliparous women. Br J Obstet Gynaecol 1998;105:1011–1017.

13 The Grit Study Group: When do obstetricians recommend delivery for a high-risk preterm growth-retarded fetus? Eur J Ostet Gyn Reprod Biol 1996;67:121–126.

14 Vossbeck S, de Camargo OK, Grab D, Bode H, Pohlandt F: Neonatal and neurodevelopmental outcome in infants born before 30 weeks of gestation with absent or reversed and –diastolic flow velocities in the umbilical artery. Eur J Paediatr 2001;160:128–134.

15 Zaw W, Gagnon R, da Silva O: The risks of adverse neonatal outcome among preterm small for gestational age infants according to neonatal versus fetal growth standards. Pediatrics 2003;111:1273–1277.

16 Piper JM, Xenakis EM, McFarland M, Elliott BD, Berkus MD, Langer O: Do growth-retarded premature infants have different rates of perinatal morbidity and mortality than appropriately grown premature infants? Obstet Gynecol 1996;87:169–174.

17 Reiss I, Landmann E, Heckmann M, Misselwitz B, Gortner L: Increased risk of bronchopulmonary dysplasia and increased mortality in very preterm infants being small for gestational age. Arch Gynecol Obstet 2003;269:40–44.

18 Regev RH, Lusky A, Dolfin T, Litmanovitz I, Arnon S, Reichman B: Excess mortality and morbidity among small-for-gestational-age premature infants: a population-based study. J Pediatr 2003;143:186–191.

19 Brion LP, Primhak RA, Ambrosio-Perez I: Diuretics acting on the distal renal tubule for preterm infants with (or developing) chronic lung disease. Cochrane Database Syst Rev 2002:CD001817.

20 Thomas W, Speer CP: Management of infants with bronchopulmonary dysplasia in Germany. Early Hum Dev 2005;81:155–163.

21 Robel-Tillig E, Knüpfer M, Vogtmann C: Cardiac adaptation in small for gestational age neonates after prenatal hemodynamic disturbances. Early Hum Dev 2003;72:123–129.

22 Bardin C, Zelkowitz P, Papageorgiou A: Outcome of small-for-gestational age and appropriate-for-gestational age infants born before 27 weeks of gestation. Pediatrics 1997;100:E4.

23 Hawdon JM, Ward Platt MP: Metabolic adaptation in small for gestational age infants. Arch Dis Child 1993;68:262–268.

24 Lucas A, Morley R, Cole TJ: Adverse neurodevelopmental outcome of moderate neonatal hypoglycaemia. Br Med J 1988;297:1304–1308.

25 Doctor BA, O'Riordan MA, Kirchner HL, Shah D, Hack M: Perinatal correlates and neonatal outcomes of small for gestational age infants born at term gestation. Am J Obstet Gynecol 2001;185:652–659.

26 Hannam S, Lees C, Edwards RJ, Greenough A: Neonatal coagulopathy in preterm small-for-gestational-age infants. Biol Neonate 2003;83:177–181.

27 Meberg A, Jakobsen E, Halvorsen K: Humoral regulation of erythropoiesis and thrombopoiesis in appropriate and small for gestational age infants. Acta Paediatr Scand 1982;71:769–773.

28 Petursson SR, Chervenick PA: Effects of hypoxia on megakaryocytopoiesis and granulopoiesis. Eur J Haematol 1987;39:267–273.

29 Carr R, Modi N, Dore C: G-CSF and GM-CSF for treating or preventing neonatal infections (review). Cochrane Library number 2007:CD003066.

30 Simchen MJ, Beiner ME, Strauss-Liviathan N, Dulitzky M, Kuint J, Mashiach S, Schiff E: Neonatal outcome in growth-restricted versus appropriately grown preterm infants. Am J Perinatol 2001;17:187–192.

31 Bernstein IM, Horbar JD, Badger GJ, Ohlsosn A, Golan A: Morbidity and mortality among very-low-birth-weight neonates with intrauterine growth restriction: The Vermont Oxford Network. Am J Obstet Gynecol 2000;182:198–206.

32 Spinillo A, Capuzzo E, Stronati M, Ometto A, De Santolo A, Acciano S: Obstetric risk factors for periventricular leukomalacia among preterm infants. Br J Onstet gynaecol 1998;105:865–871.

33 Amato M, Konrad D, Hüppi P, Donati F: Impact of prematurity and intrauterine growth retardation on neonatal hemorrhagic and ischemic brain damage. Eur Neurol 1993;33:299–303.

34 Kaiser JR, Gauss CH, Pont MM, Williams DK: Hypercapnia during the first 3 days of life is associated with severe intraventricular hemorrhage in very low birth weight infants. J Perinatol 2006;26:279–285.

35 Lindner N, et al: Risk factors for Intraventricular hemorrhage in very low birth weight premature infants: a retrospective case-control study. Pediatrics 2003;111:e590–e595.

36 Robel-Tillig E, Vogtmann C, Faber R: Postnatal intestinal disturbances in small-for-gestational-age premature infants after haemodynamic disturbances. Acta Paediatr 2000;89:324–330.

37 Müller-Egloff S, Strauss A, Spranger V, Genzel-Boroviczeny O: Does chronic prenatal Doppler pathology predict feeding difficulties in neonates? Acta Paediatr 2005;94:1632–1637.

38 McGuire W, Anthony MY: Donor human milk versus formula for preventing necrotising enterocolitis in preterm infants: systematic review. Arch Dis Child Fetal Neonatal Ed 2003;88:F11–F14.

39 Robel-Tillig E, Knüpfer M, Pulzer F, Vogtmann C: Blood flow parameters of the superior mesenteric artery as an early predictor of intestinal dysmotility in preterm infants. Pediatr Radiol 2004;34:958–962.

40 Pezzati M, Dani C, Tronchin M, Filippi L, Rossi S, Rubaltelli FF: Prediction of early tolerance to enteral feeding by measurement of superior mesenteric artery blood flow velocity: appropriate-versus small-for-gestational-age preterm infants. Acta Paediatr 2004;93:797–802.

41 Latini G, De Mitri B, Del Vecchio A, Chitano G, De Felice C, Zetterström R: Foetal growth of kidneys liver and spleen in intrauterine growth restriction: 'programming' causing 'metabolic syndrome' in adult age. Acta Paediatr 2004;93:1635–1639.

42 Hotoura E, Argyropoulou M, Papadopoulou F, Giapros V, Drougia A, Nikolopoulos P, Andronikou S: Kidney development in the first year of life in small-for-gestational-age preterm infants. Pediatr Radiol 2005;35:991–994.

43 Allegaert K, Anderson BJ, van den Anker JN, Vanhaesebrouck S, de Zegher F: Renal drug clearance in preterm neonates: relation to prenatal growth. Ther Drug Monit 2007;29:284–291.

44 Giapros V, Papadimitriou P, Challa A, Andronikou S: The effect of intrauterine growth retardation on renal function in the first two months of life. Nephrol Dial Transplant 2007;22:96–103.

Matthias Knüpfer, MD
Department of Neonatology
Hospital for Children and Adolescents, University of Leipzig
Liebigstrasse 20a
DE–04103 Leipzig (Germany)
Tel. +49 341 9726857, Fax +49 341 9723579, E-Mail matthias.knuepfer@medizin.uni-leipzig.de

Kiess W, Chernausek SD, Hokken-Koelega ACS (eds): Small for Gestational Age. Causes and Consequences.
Pediatr Adolesc Med. Basel, Karger, 2009, vol 13, pp 116–126

Management of Short Stature in Small-for-Gestational-Age Children

Jovanna Dahlgren

Göteborg Pediatric Growth Research Center, Department of Pediatrics,
Institute for Clinical Sciences, The Queen Silvia Children's Hospital,
The Sahlgrenska Academy at Göteborg University, Göteborg, Sweden

Abstract

Birth length is shown to be the most important predictor of adult height (AH). There is a 7-fold higher risk of short AH in individuals born small for gestational age (SGA), and they comprise a third of children being short. Most short SGA children show a substantial improvement in growth rate during growth hormone (GH) treatment. This is maintained to AH with a total 'gain' of 1–2 SDS in height. The majority of these children do not have classical GH deficiency, but rather a low tissue GH responsiveness. Several studies demonstrate an initial correlation between the GH dosage and the growth response although there is a large variability in the magnitude, indicating varying individual responsiveness. This may implicate an individualization of GH dosage based on GH secretion, age at start of treatment and difference to parental heights, all predictors of growth response. GH treatment is found to have a positive effect on body composition, blood pressure and lipid metabolism in short SGA children. However, as GH elevates the fasting insulin levels, it is important to evaluate the effect of treatment on carbohydrate metabolism. This is especially important, as being born SGA by itself implies increased risk to develop type 2 diabetes later in life. Copyright © 2009 S. Karger AG, Basel

Early Growth and Risk of Short Stature during Childhood

About 80–85% of children born small for gestational age (SGA) have a rapid catch-up growth during the first 12 months of life and reach a height SDS above −2 standard deviation scores (SDS). A remaining 5% will reach this cut-off at an age of 2 years, leaving 10–15% with a lack of catch-up growth [1–3]. This percentage of non catch-up growth varies between countries and seems to be increased in the Third World [4], where gastrointestinal infections and low socio-economic status have a profound influence. Another postnatal factor influencing catch-up growth is the initial extrauterine growth retardation found in SGA children born very premature due to nonoptimal substitution of supply of nutrients [5] and lack of hormones provided by the placenta. Although there is an increased risk of persistent short stature if being born premature [2], most of these children will have a slow and prolonged catch-up

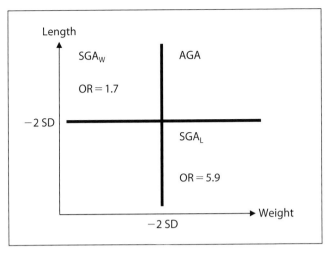

Fig. 1. Odds ratio of being short as adult in relation to birth length and birth weight. SGA$_W$ = SGA only by weight; SGA$_L$ = SGA only by length.

period compared to term SGA infants, with attained height catch-up first at an age of 4–5 years [6].

Risk of Short Stature as Adult

There is a 5- to 7-fold higher risk of short stature among adults who were born SGA, compared to those born appropriate for gestational age (AGA) [1, 2]. French data show that adult men who were born SGA are, on average 7.5 cm, and women are 9.6 cm below their mid-parent target height (TH) [7]. This is at the lower end of the range compared with other studies [1, 8], which find a growth deficit around 1 SDS below the mid-parental height (MPH).

Variables predicting lack of catch-up growth are short parental heights [1, 9, 10], chromosomal aberrations and having any syndrome [11]. Interestingly, mean MPH (expressed as SDS) in the group of short children born SGA is about 1.2 SD lower than in controls [3, 12]. However, TH seems to influence mainly childhood growth compared to pubertal growth [8].

Birth length has been shown in several population-based studies to be the single most important predictor of adult height (AH) [1, 2, 10], but also birth weight seems to be important [3]. Odds ratio for adult short stature was 5.9 if low birth length, but only 1.7 if low birth weight [2]. In other words, the shorter the newborn child, the highest risk of non-catch-up growth and if lean at birth the better postnatal catch-up growth (fig. 1). This may indicate different etiologies/entities behind the smallness at birth and therefore different growth patterns during postnatal life.

Aetiology and Risk of Short Stature – Any Link?

Short children born SGA form a heterogeneous group with various etiologies to their smallness and treatment should be preceded by an effort to identify the underlying etiology, although the cause is often not clear. Genetic abnormalities, for example trisomy 13 or 18, and mutations in the growth hormone (GH)-insulin-like growth factor (IGF) axis have been associated with small size at birth and reduced postnatal growth. These include insulin-like growth factor I (IGF-I) and IGF-I receptor gene deletions, point mutations and polymorphisms [13–16]. Moreover, intrauterine growth restriction (IUGR) due to congenital viral infection or parasite infection does not show postnatal catch-up growth [17]. Children with fetoalcohol syndrome or expose to alcohol lack also postnatal catch-up growth [18].

Rapid catch-up growth during postnatal life on the other hand, is believed to be more frequent if IUGR was a consequence of adaptation (occurs in small mothers or twin pregnancies). Also, children to mothers with placental insufficiency, insufficient maternal nutrient supply or exposure to smoking are believed to experience catch-up growth, especially by weight gain. However, the relationship between etiology of IUGR and postnatal growth pattern is not extensively delineated as published large prospective population-based studies are lacking.

Impact of Growth Retardation during the Different Trimesters of Gestation on Short Stature and Body Proportionality

Growth restriction occurring already early in gestation will lead to a permanent influence in the size of the individual. Such an example is the group of children being symmetric SGA, which show both low birth length and birth weight (SGA_{wL}) [1]. Data tell us that the fetal restriction occurred in this group during early gestation [19]. Such an example is the child with Silver-Russell syndrome [11]. Those children tend not to have a successful spontaneous postnatal catch-up growth [1].

In other words, the more prolonged the period of IUGR, the less likely the growth deficit will be recovered.

Opposite to the symmetric group are the children born asymmetric or disproportional SGA [19]. Those children are short at birth with normal birth weight (SGA_L) and are believed to experience growth retardation in mid-gestation, although some may have just familial short stature, which as expected will not lead to postnatal catch-up growth [1].

Finally, those children defined SGA by low birth weight (SGA_w) are most likely to experience fetal malnutrition late in gestation [19] and are found to have rapid postnatal catch-up growth [1]. However, the hypothesis of body proportionality at birth as an indirect marker of timing of IUGR has been questioned [20].

Growth Hormone Secretion Status in Short Small-for-Gestational-Age Children

Children born SGA comprise one-third of children who are short during childhood [1], thus being a substantial part of short children presenting at pediatric growth clinics. Interestingly, a large study has shown that neither circulating concentrations of GH, IGF-I, nor insulin-like growth factor binding protein-3 (IGFBP-3) during infancy are predictive of subsequent growth [21].

Short children born SGA are usually not classical GH deficient (GHD), but have been found to have either low GH secretion or reduced sensitivity to GH [22–25]. As many as 50–60% investigated children born SGA with short stature have either 24-hour GH profile abnormalities or subnormal responses to arginine provocation with reduced plasma IGF-I and IGF-II levels, indicating GH insufficiency [22]. Moreover, Boguszewski and coworkers reported that short children born SGA have both lower mean spontaneous GH secretion rates [25] and lower serum IGF-I values [26] than children born AGA. They also found that the GH secretion rate of 24-hour GH profiles was the strongest predictor (a negative correlation) of the short-term growth response on GH treatment, followed by IGF-I and leptin levels at investigation previous to treatment. However, whether measures of GH secretion provide clinically useful information for GH treatment and the subsequent growth response on treatment of short SGA children is controversial [27–29].

Effects of Growth Hormone Therapy on Prepubertal Growth in Short Small-for-Gestational-Age Children

Despite their GH secretion status at investigation, the majority of short children born SGA show improvement in growth rate during GH treatment. Nevertheless, an endocrine evaluation of a presenting short child born SGA is recommended to exclude not only GHD but also hypothyroidism, celiac disease, malnutrition, renal failure or chronic inflammation.

The improvement in growth rate is substantial if treatment is begun at an early age [12, 30–32, 35]. Average height gains after 3 years of GH treatment range from 1.2 to 2.0 SD for doses of 35–70 μg/kg/day [30]. A progressive decrease in the effect of GH treatment on growth rate occurs over time [28, 33]. Interestingly and of clinical relevance is the finding that the first-year response on GH is a strong predictor of further height gain [35].

Factors favorable for substantial growth response during the first 2–3 years on GH treatment are young age and low height SDS at start [27]. Prepubertal years on GH treatment [12] and prepubertal GH dose [30–32, 36] are the main factors for height normalization. Interestingly, results from several studies suggest that the cumulative GH dose received rather than the dosing regimen determines the growth response during the prepubertal years [28, 29, 34].

Whether TH is important factor for catch-up growth on GH treatment is debated. Some studies have found such correlation [12, 27, 29, 35], while others did not

Table 1. Results of published data – variables predicting short-term (0.5–6 years) height gain

Dahlgren and Albertsson Wikland [12]	Boguszewski et al. [26]	Sas et al. [27]	de Zegher et al. [32]	Ranke et al. [35]	van Dijk et al. [36]
Dependent variable: height gain					
Years on GH			GH dose	GH dose	GH dose
IGF-I SDS	IGF-I SDS				IGF-I SDS
	delta IGF-I	bone age			GH status
Age at start	age at start	age at start	age at start	age at start	age at start
Height at start			height at start		height at start
Weight at start			weight at start	weight at start	
Diff. at start	mothers height			MPH	

Diff. = Difference between height at start GH treatment and mid-parental height (MPH).

[27, 36, 37]. For a review of the predictive factors for catch-up growth on treatment during a short-term period (0.5–6 years on GH treatment), see table 1.

Change in Insulin-Like Growth Factor I during Growth Hormone Treatment

GH treatment in most SGA children is not substitutive but is pharmacological and will lead to a rapid increase in IGF-I and IGFBP-3 levels [26, 28]. The short-term change in IGF-I levels expressed as SDS during the first months and year of GH treatment correlate with the subsequent growth response [26]. Moreover, IGF-I SDS levels at start have been shown to correlate with prepubertal height gain [12]. However, a recently published study showed that the free IGF-I levels but not the total IGF-I correlate with growth response during GH treatment [38].

Effects of Growth Hormone Therapy on Pubertal Growth

Normal pubertal growth pattern without GH treatment and influencing parameters on this pattern are only partially studied. GH therapy in short children born SGA seems not to influence on the age at onset and duration of puberty, which is about 4 years from onset of puberty to AH in girls and 5 years in boys [29, 34].

In contrast to short children born AGA, who have a gain of 0.6–0.7 SDS in height during puberty, short children born SGA keep their prepubertal height SDS to AH without GH treatment [8, 12]. Pubertal growth on GH treatment seems to be less influenced by GH dose [12, 29] than during the prepubertal years. The gain in height achieved before puberty on GH treatment normalizes height during childhood in most children born SGA and is either maintained or in some degree lost through puberty to AH [12, 29], but found to correspond to the height of their parents [12]. In other words, normalization of height before puberty is essential.

Interestingly, short children born SGA with partial GHD have a better height gain during pubertal years than those non-GHD children [12]. Moreover, during pubertal years GH status has an impact on height gain [12], which could be speculated to be due to a substantial decrease in IGF-I SDS levels during pubertal years on GH treatment in non-GHD children [unpubl. data]. However, the reasons for these findings merit further evaluation.

Little is known about the start of GH treatment in short SGA children near pubertal onset or during pubertal years, but it seems that very little gain in AH is achieved in individuals with such treatment [7, 12, 37, 39]. This is illustrated with the data from a French SGA study [37], where the gain in height was only of 0.6 SDS if treatment started from a mean age of 12.7 years (compared to a gain of 2.2 SDS if treatment started at the age of 7 years in a Dutch study [29]).

Effect of Growth Hormone Therapy on Total Growth from Start to Adult Height

Age at start of GH treatment, the prepubertal years on GH treatment and prepubertal dose are the strongest variables correlating with height gain to AH [12, 29–31]. Other significant variables have also been published (see table 2 for more details).

Several contradictory results are published concerning the magnitude of the gain in height in short children born SGA but these results could be explained by differences in study design. The main findings can be summarized as follows: If therapy is started at an early age [12, 29] this gain is of a large magnitude (1.7–2.1 SDS), but if this is not the case the treatment will lead to a less overall height gain (0.6–0.9 SDS) [12, 37, 39].

Effect of Growth Hormone Therapy on Adult Height

GH treatment improves not only short-term growth but also AH [7, 12, 29, 37]. 85% of short children born SGA treated with GH for a long-term period will reach normal AH [12, 29]. As the mean TH is lower than the normal population, this means that 98% of these children reach an AH within their TH [29]. The variables predicting AH are TH, height SDS and age at start GH treatment [12, 29]. GH status is not found to predict AH [12, 29].

It is important to have in mind that adult height and gain to adult height are completely different dependent variables to study and therefore not surprising that the explaining variables will diverge. For comparisons, see table 2. Unfortunately, not all studies have investigated the correlations to gain, but just to the attained AH [36].

Table 2. Results of published data – variables predicting adult height and height gain from start to adult height (delta height).

	Dahlgren and Albertsson Wikland [12]	Van Pareren et al. [29]	Carel et al. [37]	Cautant et al. [39]
GH dose used per day	0.1–0.2 U/kg	0.1–0.2 U/kg	0.2 U/kg	0.03–0.1 U/kg
Magnitude of delta height	0.9 and 1.7 SDS, respectively	1.8 and 2.1 SDS, respectively	0.6 SDS	0.6 SDS
Dependent variable: adult height	father's height height at start age at start GH_{max} AITT weight at start	TH height at start bone age retardation pretreatment HV	duration of GH height at start bone age retardation	TH age at start BMI at start
Dependent variable: delta height	duration of GH age at start height at start weight at start Diff. SDS at start GH_{max} AITT	age bone age		

Diff. = Difference between height at start GH treatment and MPH; GH_{max} AITT = maximum GH peak during arginin insulin tolerance test; TH = target height; HV = height velocity.

However, as an example to mention is the impact of height at start on the AH. Not surprisingly, the taller the child at start of treatment the better AH. On the contrary to this, the larger difference to TH the better the height gain will be.

Potential Risks of Therapy – Supraphysiological Circulating Insulin-Like Growth Factor I

High-dose GH treatment in short SGA children may lead to high serum GH and IGF-I levels [36, 38] and therefore it is recommended regularly monitoring of IGF-I levels to ensure that these remain within the normal range.

Metabolic Effects of Growth Hormone Therapy in Short Small-for-Gestational-Age Children

GH treatment has anabolic effects with increase of fasting insulin levels and decreased insulin sensitivity, as well as it enhances lipolysis and gluconeogenesis.

In other words, the benefits of GH therapy are not just in terms of height, but also in body composition, blood pressure and lipid levels. This is also the case in short children born SGA [40, 41].

However, there is a concern of GH treatment worsening the risk of developing the metabolic syndrome found in subjects born SGA already without GH treatment. Increased risk of metabolic syndrome – type 2 diabetes or insulin resistance, hypertension and hyperlipidemia – has been reported in adults who were born SGA [42], especially if catch-up growth by weight.

As many as 8% of investigated short children born SGA are found to have impaired glucose tolerance before GH treatment started [41]. During treatment no worsening was seen, but an increase of fasting insulin has been shown [41]. Discontinuation at AH, after fulfilled GH treatment, normalizes GH-induced insulin insensitivity. Encouraging, discontinuation after long-term treatment seems not to change the beneficial effect of GH on blood pressure and blood lipids [43].

Discontinuation of GH Treatment

It has been explored and published the effects of continuous and/or discontinuous regimens of GH [34]. These data found both regimens to be effective growth promoting. However, a reduction of height velocity and height SDS is seen after interruption of GH therapy [34, 44]. Therefore, it is recommended long-term continuous GH treatment with standard dose, especially if treatment starts several years before start of puberty. However, discontinuous high dose treatment may be chosen if the primary objective is rapid normalization of height in younger age [28, 30, 34].

Consensus on Indications of Growth Hormone Therapy in Short Small-for-Gestational-Age Children

The optimal GH dose in short SGA children is currently debated, which is mentioned in a recent consensus paper [30]. Approved indication of GH treatment in short children born SGA varies over the continents. Since 2001, the Food and Drug Administration in the USA has given its approval to treat with recombinant GH treatment if the child lacks catch-up growth at an age of 2 years. The recommended dose here is 70 μg/kg/day. In Europe, on the other hand, the child has to be 4 years of age, below −2.5 SDS in height at start, lack catch-up growth and have a distance in height SDS to mid-parental height of more than one SDS. The permitted dosage is instead 35 μg/kg/day.

To summarize, the recommended GH regimens differ between USA and Europe with respect to GH dosage, age at which therapy is allowed to start and the degree of short stature required.

Future Approaches to Growth Hormone Treatment

Validated prediction models giving an index of responsiveness for each short child are useful for individualizing the dose of GH [45]. This is also the case in short children born SGA [35], a heterogeneous group where children either may have a low or a normal/high spontaneous GH secretion level. Therefore, it is crucial to find the optimal hormonal treatment to their shortness and pretreatment GH levels may point out those with high or low responsiveness to GH [45]. Based on these pretreatment evaluations, it may be decided that a child either should receive normal GH dose or high dose (alternatively combined with IGF-I treatment if low GH responsiveness).

Moreover, IGF-I monitoring is another approach to GH responsiveness and should not only be used as a safety matter but could in the future also serve as a tool for dose optimization.

Conclusion

Long-term, continuous GH treatment in short children born SGA leads to a normalization of height during childhood and to adult age. GH treatment should start at an early age as younger, shorter and lighter children at the start of GH treatment have better growth response, are taller at onset of puberty and achieve better AH. GH therapy is not only efficient in normalizing height but is also safe in these children. We recommend early surveillance in a growth clinic for those children born SGA with lack of catch-up. Moreover, regular long-term surveillance of those short SGA children who receive GH with measurement of glucose homeostatis, free IGF-I, lipid levels and blood pressure is essential.

References

1   Karlberg J, Albertsson-Wikland K: Growth in full-term small-for-gestational-age infants: from birth to final height. Pediatr Res 1995;38:733–739.
2   Tuvemo T, Cnattingius S, Jonsson B: Prediction of male adult stature using anthropometric data at birth: a nationwide population-based study. Pediatr Res 1999;46:491–495.
3   Hokken-Koelega AC, De Ridder MA, Lemmen RJ, Den Hartog H, De Muinck Keizer-Schrama SM, Drop SL: Children born small for gestational age; do they catch up? Pediat Res 1995;38:267–271.
4   Hofvander Y: International comparisons of postnatal growth of low birthweight infants with special reference to differences between developing and affluent countries. Acta Paediat Scand 1982;296 (suppl):14–18.
5   Niklasson A, Engstrom E, Hard AL, Wikland KA, Hellstrom A: Growth in very preterm children: a longitudinal study. Pediatr Res 2003;54:899–905.
6   Gibson AT, Carney S, Cavazzoni E, Wales JK: Neonatal and postnatal growth. Horm Res 2000;53 (suppl 1):42–49.
7   Chaussain J, Colle M, Ducret J: Adult height in children with prepubertal short stature secondary to intrauterine growth retardation. Acta Pediatr 1994; 399(suppl):72–73.
8   Luo ZC, Albertsson-Wikland K, Karlberg J: Length and body mass index at birth and target height influences on patterns of postnatal growth in children born small for gestational age. Pediatrics 1998;102:e72.

9 Albertsson-Wikland K, Boguszewski M, Karlberg J: Children born small-for-gestational age: postnatal growth and hormonal status. Horm Res 1998;49 (suppl 2):7–13.

10 Léger J, Levy-Marchal C, Bloch J, Pinet A, Chevenne D, Porquet D, Collin D, Czernichow P: Reduced final height and indications for early development of insulin resistance in a 20 year old population born small for gestational age: regional cohort study. BMJ 1997;315:341–347.

11 Tanner JM, Lejarraga H, Cameron N: The natural history of the Silver-Russell syndrome: a longitudinal study of thirty-nine cases. Pediatr Res 1975;9: 611–623.

12 Dahlgren J, Albertsson Wikland K: Final height in short children born small for gestational age treated with growth hormone. Pediatr Res 2004;57:216–222.

13 Woods KA, Camacho-Hubner C, Savage MO, Clark AJ: Intrauterine growth retardation and postnatal growth failure associated with deletion of the insulin-like growth factor I gene. N Engl J Med 1996; 335:1363–1367.

14 Abuzzahab MJ, Schneider A, Goddard A, Grigorescu F, Lautier C, Keller E, Kiess W, Klammt J, Kratzsch J, Osgood D, Pfaffle R, Raile K, Seidel B, Smith RJ, Chernausek SD, Intrauterine Growth Retardation (IUGR) Study Group: IGF-I receptor mutations resulting in intrauterine and postnatal growth retardation. N Engl J Med 2003;349:2211–2222.

15 Vaessen N, Janssen JA, Heutink P, Hofman A, Lamberts SW, Oostra BA, Pols HA, van Duijn CM: Association between genetic variation in the gene for insulin-like growth factor-I and low birthweight. Lancet 2002;359:1036–1037.

16 Arens N, Johnston L, Hokken-Kolega A: Polymorphism in the IGF-I gene: clinical relevance for short children born small for gestational age (SGA). J Clin Endocrinol Metab 2002;87:2720.

17 Dreyfuss ML, Msamanga GI, Spiegelman D, Hunter DJ, Urassa EJ, Hertzmark E, Fawzi WW: Determinants of low birth weight among HIV-infected pregnant women in Tanzania. Am J Clin Nutr 2001;74:814–826.

18 Tran TD, Cronise K, Marino MD, Jenkins WJ, Kelly SJ: Critical periods for the effects of alcohol exposure on brain weight, body weight, activity and investigation. Behav Brain Res 2000;116:99–110.

19 Villar J, Belizan JM: The timing factor in the pathophysiology of intrauterine growth retardation syndrome. Obstet Gynecol Surv 1982;37:499–506.

20 Kramer MS, Oliver M, McLean FH, Dougherty GE, Willis DM, Usher RH: Determinants of fetal growth and body proportionality. Pediatrics 1990;86:18–26.

21 Leger J, Noel M, Limal JM, Czernichow P: Growth factors and intrauterine growth retardation. II. Serum growth hormone, insulin-like growth factor (IGF) I, and IGF-binding protein 3 levels in children with intrauterine growth retardation compared with normal control subjects: prospective study from birth to two years of age. Study Group of IUGR. Pediatr Res 1996;40:101–107.

22 de Waal WJ, Hokken-Koelega AC, Stijnen T, de Muinck Keizer-Schrama SM, Drop SL: Endogenous and stimulated GH secretion, urinary GH excretion, and plasma IGF-I and IGF-II levels in prepubertal children with short stature after intrauterine growth retardation. The Dutch Working Group on Growth Hormone. Clin Endocrinol (Oxf) 1994;41: 621–630.

23 Albertsson-Wikland K, Swedish Paediatric Study Group for Growth Hormone Treatment: Growth hormone secretion and growth hormone treatment in children with intrauterine growth retardation. Acta Paediatr Scand 1989;349(suppl):35–41.

24 Stanhope R, Ackland F, Hamill G, Clayton J, Jones J, Preece MA: Physiological growth hormone secretion and response to growth hormone treatment in children with short stature and intrauterine growth retardation. Acta Paediatr Scand 1989;349(suppl): 47–52.

25 Boguszewski M, Rosberg S, AlbertssonWikland K: Spontaneous 24hour growth hormone profiles in prepubertal small for gestational age children. J Clin Endocrinol Metab 1995;80:2599–2606.

26 Boguszewski M, Jansson C, Rosberg S, Albertsson Wikland K: Changes in serum insulin-like growth factor I (IGF-I) and IGF-binding protein-3 levels during growth hormone treatment in prepubertal short children born small for gestational age. J Clin Endocrinol Metab 1996;81:3902–3908.

27 Sas T, de Waal W, Mulder P, Houdijk M, Jansen M, Reeser M, Hokken-Koelega A: Growth hormone treatment in children with short stature born small for gestational age: 5-year results of a randomised, double-blind, dose-response trial. J Clin Endocrinol Metab 1999;84:3064–3070.

28 Lee PA, Chernausek SD, Hokken-Koelega AC, Czernichow P: International Small for Gestational Age Advisory Board. International Small for Gestational Age Advisory Board consensus development conference statement: management of short children born small for gestational age, April 24–October 1, 2001. Pediatrics 2003;111:1253–1261.

29 van Pareren Y, Mulder P, Houdijk M, Jansen M, Reeser M, Hokken-Koelega A: Adult height after long-term, continuous GH treatment in short children born SGA: results of a randomized, double-blind, dose-response GH trial. J Clin Endocrinol Metab 2003;88:3584–3590.

30 Clayton PE, Cianfarani S, Czernichow P, Johansson G, Rapaport R, Rogol A: Management of the child born small for gestational age through to adulthood: A consensus statement of the international societies of pediatric endocrinology and the Growth Hormone Research Sociaty. J Clin Endocrinol Metab 2007;92:804–810.

31 Boguszewski M, Albertsson-Wikland K, Aronsson S, Gustafsson J, Hagenäs L, Westgren U, Westphal O, Lipsanen-Nyman M, Sipilä I, Gellert P, Muller J, Madsen B: Growth hormone treatment of short children born small-for-gestational-age: the Nordic multicenter trial. Acta Paediatr 1998;87:257–263.

32 de Zegher F, Albertsson Wikland K, Wilton P, Chatelain P, Jonsson B, Lofstrom A, Butenandt O, Chaussain J-L: Growth hormone treatment of short children born small for gestational age: metaanalysis of four independent, randomised, controlled, multicentre studies. Acta Paediatr 1996;417(suppl): 27–31.

33 de Zegher F, Hokken-Koelega A: Growth hormone therapy for children born small for gestational age: height gain is less dose dependent over the long term than over the short term. Pediatrics 2005;115: e458–e462.

34 de Zegher F, Albertsson Wikland K, Wollmann HA, Chatelain P, Chaussain JL, Lofstrom A, Jonsson B, Rosenfeld RG: Growth hormone treatment of short children born small for gestational age: growth responses with continuous and discontinuous regimens over 6 years. J Clin Endocrinol Metab 2000; 85:2816–2821.

35 Ranke MB, Lindberg A, Cowell CT, Albertsson Wikland K, Reiter EO, Wilton P, Price DA: Prediction of response to GH treatment in short children born SGA: Analysis of data from KIGS. J Clin Endocrinol Metab 2003;88:125–131.

36 van Dijk M, Mulder P, Houdijk M, Mulder J, Noordam K, Odink RJ, Rongen-Westerlaken C, Voorhoeve P, Waelkens J, Stokvis-Brantsma J, Hokken-Koelega A: High serum levels of growth hormone (GH) and insulin-like growth factor-I (IGF-I) during high-dose GH treatment in short children born small for gestational age. J Clin Endocrinol Metab 2007;92:160–165.

37 Carel J-C, Chatelain P, Rochiccioli P, Chaussain J-L: Improvement in adult height after GH treatment in adolescents with short stature born SGA: Results of a randomized controlled study. J Clin Endocrinol Metab 2003;88:1587–1593.

38 Bannink EM, van Doorn J, Mulder PG, Hokken-Koelega AC: Free/dissociable IGF-I, not total IGF-I correlates with growth response during growth hormone treatment in children born small for gestational age. J Clin Endocrinol Metab 2007;May 15, epubl.

39 Coutant R, Carel JC, Letrait M, Bouvattier C, Chatelain P, Coste J, Chaussain JL: Short stature associated with intrauterine growth retardation: final height of untreated and growth hormone-treated children. J Clin Endocrinol Metab 1998; 83:1070–1074.

40 Sas T, Mulder P, Hokken-Koelega A: Body composition, blood pressure, and lipid metabolism before and during long-term growth hormone (GH) treatment in children with short stature born small for gestational age either with or without GH deficiency. J Clin Endocrinol Metab 2000;85:3786–3792.

41 Sas T, Mulder P, Aanstoot HJ, Houdijk M, Jansen M, Reeser M, Hokken-Koelega A: Carbohydrate metabolism during long-term growth hormone treatment in children with short stature born small for gestational age. Clin Endocrinol (Oxf) 2001;54:243–251.

42 Barker DJP: Fetal and Infant Origins of Adult Disease. BMJ Publishing group, London, 1991.

43 van Pareren Y, Mulder P, Houdijk M, Jansen M, Reeser M, Hokken-Koelega A: Effect of discontinuation of growth hormone treatment on risk factors for cardiovascular disease in adolescents born small for gestational age. J Clin Endocrinol Metab 2003; 88:347–353.

44 Fjellestad-Paulsen A, Simon D, Czernichow P: Short children born small for gestational age and treated with growth hormone for three years have an important catch-down five years after discontinuation of treatment. J Clin Endocrinol Metab 2004; 89:1234–1239.

45 Albertsson Wikland K, Kriström B, Rosberg S, Svensson B, Nierop A: Validated multivariate models predicting the growth response to GH treatment in individual short children with a broad range in GH secretion capacities. Pediatr Res 2000;48:475–484.

Jovanna Dahlgren, MD, PhD, Assoc. Professor
Göteborg Pediatric Growth Research Center, Department of Pediatrics, Institute for Clinical Sciences
The Queen Silvia Children's Hospital, The Sahlgrenska Academy at Göteborg University
SE–416 85 Göteborg (Sweden)
Tel. +46 702 05 88 96, Fax +46 31 84 89 52, E-Mail jovanna.dahlgren@vgregion.se

Kiess W, Chernausek SD, Hokken-Koelega ACS (eds): Small for Gestational Age. Causes and Consequences.
Pediatr Adolesc Med. Basel, Karger, 2009, vol 13, pp 127–133

Puberty and Adrenarche in Small-for-Gestational-Age Children

Anita C.S. Hokken-Koelega

Division of Endocrinology, Department of Paediatrics, Sophia Children's Hospital,
Erasmus University Medical Center, Rotterdam, The Netherlands

Abstract

Most children born small for gestational age (SGA) have a normal adrenarche and pubarche and a normal timing of pubertal development. Onset of puberty occurs at a normal age, but relatively early within the normal range and in some children unexpectedly early for their short stature. Some studies indicated that SGA boys and girls have a slightly reduced pubertal growth spurt, which leads to a shorter adult stature. In Northern Spanish girls an association was found between precocious pubarche, hyperandrogenism, hyperinsulinemia, polycystic ovaries syndrome and lower birth weight, but this could not be confirmed by several studies in children born SGA. Differences in ethnic background, nutrition, different definitions and other yet unknown variables may play a role in the reported variations in timing of adrenarche and puberty.   Copyright © 2009 S. Karger AG, Basel

Postnatal growth studies in children born small for gestational age (SGA) demonstrated that the majority (90%) exhibit catch-up growth to a normal height within the first 2–3 years after birth [1, 2]. The remainder of these children tend to grow parallel to, but below the normal height centiles and will have persistent short stature.

Studies on the timing of adrenarche, puberty and pubertal growth in children born SGA are limited. Most studies investigated children born SGA, without any distinction between children who experienced catch-up growth to a normal height and those who remained short.

This chapter reviews current knowledge and concepts about the influence of fetal growth restriction and postnatal weight gain on the timing of adrenarche and puberty in the general population. Subsequently, it presents epidemiological data on the timing of puberty of children born SGA compared to the general population, followed by clinical data on adrenarche and puberty in short and normal statured children born SGA and on puberty in girls with precocious pubarche who were born with a low birth weight.

Current Concepts about Prenatal and Postnatal Influences on Timing of Puberty

Prenatal Growth Restraint

During prenatal life, there are critical phases, so-called critical windows of time, during which tissues are particularly sensitive to reprogramming [3]. When fetal growth retardation occurs during these critical windows, the fetus will alter its metabolic and endocrine set points in order to survive [4]. A transient prenatal growth constraint may thus be followed by a permanent resetting of endocrine axes. This may also have an impact on the timing of puberty. The alteration of the endocrine set points may occur due to changes in fetal environment but it may also be the result of a genetic predisposition in combination with certain fetal environmental factors.

Postnatal Weight Gain

Recently, there is increasing evidence that not only prenatal growth but also infancy weight gain may be a critical determinant of the timing of puberty [5, 6]. In a large longitudinal cohort study, growth and weight gain during the first 2 years of life predicted earlier puberty [7]. Other studies found an association between weight gain during the first 6 months of life and the timing of puberty [8]. These findings suggest that the programming of pubertal development may also occur during critical windows of time after birth, particularly during the first 2 years. Variations in nutrition may programme subsequent growth and development.

Combination of Prenatal Growth Restraint and Postnatal Weight Gain

Particularly the newborn with a relatively low birth weight, which has experienced nutritional restraint in utero, will show infancy catch-up weight gain if postnatal nutrition is normal. The rapid transition from the prenatal inadequate nutritional environment to the postnatal adequate nutritional environment may be the signal for acceleration of tempo of growth, and early induction of puberty to maximize the potential for reproduction. Engelbregt et al. [9] assessed the effects of intrauterine malnutrition during late gestation in comparison with postnatal food restriction on the onset of puberty in male and female rats. They showed that both prenatal and postnatal malnutrition changed the endocrine programming and consequently the onset and timing of puberty. This might also influence the timing of adrenarche, as higher levels of andrenal androgens, dehydroepiandrosterone (DHEA) and dehydroepiandrosterone sulphate (DHEAS) were found in 8-year-old children who were born with a relative low birth weight and who had experienced rapid infant weight gain [10]. These effects could be mediated through epigenetic effects of pregnancy or early postnatal nutrition on genes which regulate appetite and fat deposition, or alternatively through a genetic predisposition to rapid infancy weight gain. Thus, early determinants of size at birth and rapid postnatal catch-up in weight seem to be important determinants of long-term health and the timing of puberty.

The mechanisms whereby changes in infant weight gain might not only influence tempo of growth but also the timing of pubertal development are largely unknown. In the general population, rapid and increased childhood weight gain with increased (abdominal) fat acquisition has been related to earlier pubertal maturation. Also, children with early puberty have a higher weight/height ratio [11]. The transition from a relatively low birth weight to normal BMI during childhood has been associated with a higher percentage of body fat, increased central fat deposition and reduced lean tissue in children and adults [12–16]. Several studies demonstrated that particularly faster weight gain during the first months of life predicted a greater percentage body fat at later age [17, 18]. The transition from a relatively low birth weight to a greater percentage body fat at later age has also been associated with an increased risk of insulin resistance [19, 20]. Based on these data, one may speculate that the relative higher fat mass after rapid postnatal weight gain induces insulin resistance, thereby triggering other hormonal changes that may affect the tempo of growth and the timing of pubertal development [21].

Epidemiological Data on Adrenarche and Timing of Puberty in Children Born Small for Gestational Age

Epidemiological studies could not confirm that children born SGA have a significantly different adrenarche and puberty compared to the general population. Most studies did not specify their data for children who caught up to a normal stature and those who remained short. The Swedish studies made a distinction between children born light versus born short or both light and short.

Persson et al. [22] studied a large group of children born as singletons in Uppsala in 1973–1977 who were followed from birth until 16 years of age. Boys born light or SGA were 4 cm shorter at puberty onset than boys born appropriate for gestational age (AGA). Among girls, similar differences in height between groups were found. Prematurely born children were equally as tall at puberty onset as children in the reference cohort. For boys, the mean (SD) age at puberty onset did not differ between boys born with a weight or length below -2 SDS compared to that of boys born AGA. Boys started puberty at a mean (SD) age of 12.1 (1.1) years. Girls born light or short for gestational age were in comparison with girls born AGA 5 months younger at the onset of puberty (mean (SD) age 10.6 (1.2) vs. 11.1 (1.0) years) and 4 months younger at menarche (mean (SD) age 12.7(1.1) vs. 13.1 (1.0) years, respectively). Only girls born light for GA had an earlier onset of puberty and menarche.

Another population-based study involved 3650 healthy Swedish individuals born term [23]. Within this cohort, 111 (3.0%) individuals had a low birth weight (< -2 SDS), 141 (3.9%) had short birth length (< -2 SDS), 54 (1.5%) had both low birth weight and -length. It was concluded that children who showed full catch-up growth in height attained puberty at a normal or early age and reached a mean adult height of -0.7 SDS. Children who remained short throughout childhood had a puberty onset at a 'relatively early age'. Their mean adult height was -1.7 SDS. No exact data on mean age at onset of puberty were given.

A French study in 236 full-term singleton subjects born SGA (birth length and/or weight < P3) selected from a population-based registry showed that puberty in girls occurred at a normal age as their mean age of menarche was comparable with that of an appropriate for gestational age (AGA) population [24]. A subsequent study in the same group demonstrated that the difference between adult height SDS and the pre-pubertal height SDS was similar for the SGA and AGA groups suggesting that adult height in the SGA population was not influenced by puberty and that pubertal growth spurt was normal in SGA.

Although none of the studies reported puberty into great detail, authors do seem to agree that puberty in children born SGA starts at a normal age, at a relatively early age within the normal range and unexpectedly early for their short stature. None of them reported precocious puberty in children born SGA.

Clinical Data on Adrenarche and Puberty in Children Born Small for Gestational Age

Adrenarche and Pubarche

Boonstra et al. [25] showed that the serum DHEAS levels of a large group of 185 Dutch short children born SGA (3–9 years) were comparable with those of normal children of the same age and weight. There was no relation between the serum DHEAS levels and birth weight SDS or birth length SDS. The incidence of precocious pubarche was comparable with controls. Jacquet et al. [26] reported that in a cohort of young women born SGA no increased levels of androgens were found.

Thus, onset of adrenarche of children born SGA is not different compared to the general population.

Timing of Pubertal Development

Timing of puberty was evaluated in a large group of 75 short Dutch children born SGA (birth length < −2 SDS for GA) [27]. The age at onset of puberty and duration of puberty in boys and girls and age at menarche in girls were comparable with children born appropriate for GA with a normal height (AGA; birth length > −2 SDS0).

Lazar et al. [28] compared timing of puberty of 76 short children born SGA (height SDS< −1.7) with that of 52 children born AGA with a normal height. Menarche occurred in the SGA girls within the normal age range but 6 months earlier than in AGA girls (12.6 vs. 13.0 years; p < 0.01). Although onset and duration of puberty were similar, the short SGA children exhibited an earlier peak height velocity at Tanner stage 2–3, followed by a decelerated growth and earlier fusion of epiphyses. The authors concluded that short children born SGA have a normal pubertal course; however, with a distinct pubertal growth pattern.

Hernàndez et al. [29] investigated in a longitudinal study, the pubertal course of 30 girls born SGA with a normal height after postnatal catch-up in height (height SDS −0.52) in comparison with 35 matched girls born AGA (height −0.42). Children were recruited from the community between age 7 and 10 years. SGA girls had a similar age at onset of pubarche and puberty and similar pubertal development and ovarian/uterus measurements by ultrasound. There was only a slightly higher serum estradiol level at the onset of puberty.

Thus, clinical studies show that most children born SGA have a normal timing of puberty compared to the general population.

Puberty in Girls Who Presented with Precocious Pubarche or Early Puberty

Ibanez et al. [30] reported in Northern Spanish girls with precocious pubarche, a significantly lower birth weight than normal and ovarian hyperandrogenism. They also found an exaggerated adrenarche and hyperinsulism in a group of adolescent girls born with a lower birth weight < −1.5 SD [31]. Another study showed that non-obese girls with precocious pubarche had excess body fat and particularly central fat mass throughout all pubertal stages, whereas increased central fat was related to hyperinsulinemia and hyperandrogenemia [32]. This pointed to a possible association between fetal growth and the occurrence of precocious adrenarche, pubarche, polycystic ovaries syndrome (PCOS) and hyperinsulinism [33]. The authors concluded that early weight gain and insulin resistance have a central role in the precocious pubarche as well as the earlier pubertal development [34]. Insulin sensitization using metformin treatment resulted in a significant reduction in serum adrenal androgen levels and a slower progression towards menarche and a lower risk for PCOS [35].

However, the frequency of precocious pubarche, earlier puberty and PCOS in the Spanish girls might be related to their ethnic background. These findings remain to be confirmed in larger populations of SGA children with different ethnicities.

Gonadal Development

There are no substantial data to support gonadal dysfunction, reduced fertility or early menopause in female subjects born SGA. Reduced uterine and ovarian sizes were found in adolescent Northern Spanish girls born SGA, with increased secretion of adrenal and ovarian androgens and hyperinsulinemia [36].

Conclusions

Current concepts, based on data in the general population, suggest that children with a lower birth weight and rapid postnatal weight gain are at a higher risk for more fat, reduced insulin sensitivity and a different timing of puberty. In clinical practice, however, most children born SGA show a normal adrenarche and pubarche and normal timing of puberty, indicating that the reported prenatal and postnatal influences are quite subtle. Puberty in children born SGA starts at a normal age, albeit relatively early within the normal range and in some children unexpectedly early for their short stature. Some studies in boys and girls indicated that pubertal growth spurt is slightly reduced, which may result in a shorter adult stature than predicted at a younger age. In Northern Spanish girls, an association was found between precocious pubarche, hyperandrogenism, hyperinsulinemia, PCOS and lower birth weight. However, this was not found in children born SGA living in other countries. These differences may be related to differences in ethnic background, nutrition and other yet unknown variables.

References

1 Albertson-Wikland K, Karlberg J: Natural growth in children born small for gestational age with and without catch-up growth. Acta Paediatr Scand 1994;399(suppl):64–70.
2 Hokken-Koelega A, de Ridder M, van Lemmen R, den Hartog H, de Muinck Keizer-Schrama S, Drop S: Children born small for gestational age: do they catch up? Pediatr Res 1995;38:267–271.
3 Widdowson EM, McCance RA: A review: new thoughts on growth. Pediatr Res 1975;9:154–156.
4 Barker DJ: Intrauterine programming of adult disease. Molecular Medicine Today 1995;1:418–423.
5 Cooper C, Kuh D, Egger P, Wadsworth ME, Barker D: Childhood growth and age at menarche. Br J Obstet Gynaecol 1996;103:814–817.
6 Lou ZC, Karlberg J: Critical growth phases for adult shortness. Am J Epidemiol 2000;15:125–131.
7 Dos Santos Silva I, De Stavola BL, Mann V, Kuh V, Hardy R, Wadsworth ME: Prenatal factors, childhood growth trajectories and age at menarche. Int J Epidemiol 2002;31:405–412.
8 Adair LS: Size at birth predicts age at menarche. Pediatrics 2001;107:E59.
9 Engelbregt MJT, Houdijk MEC, Popp-Snijders C, Delemarre-van de Waal HA: The effects of intrauterine growth retardation and postnatal undernutrition on onset of puberty in male and female rats. Pediatr Res 2000;48:803–807.
10 Ong KK, Potau N, Petry CJ, Jones R, Ness AR, Honour JW, De Zegher F, Ibanez L, Dunger DB: Opposing influences of prenatal and postnatal weight gain on adrenarche in normal boys and girls. J Clin Endocrinol Metab 2004;89:2647–2651.
11 Cameron N, Demerath EW: Critical periods in human growth and their relationships to diseases of aging. Am J Phys Anthropol 2002;35:159–184.

12 Law CM, Barker DJ, Osmond C, Fall CH, Simmonds SJ: Early growth and abdominal fatness in adult life. J Epidemiol Community Health 1992;3: 184–186.
13 Hediger ML, Overpeck MD, Maurer KR, RJ Kuczmarski, McGlynn A, Davis WW: Growth of infants and young children born small or large for gestational age: Findings from the Third National Health and Nutrition Examination Survey, 1998, vol 152, pp 1225–1231.
14 Loos RJ, Beunen G, Fagard R, Derom C, Vlietinck R: Birth weight and body composition in young adult men: a prospective twin study. Int J Obes Relat Metab Disord 2001;25:1537–1545.
15 Loos RJ, Beunen G, Fagard R, Derom C, Vlietinck R: Birth weight and body composition in young adult women: a prospective twin study. Int J Obes Relat Metab Disord 2002;75:676–682.
16 Garnett SP, Cowell CT, Baur LA, Fay RA, Lee J, Coakley J: Abdominal fat and birth size in healthy prepubertal children. Int J Obes Relat Metab Disord 2001;25:1667–1673.
17 Ong KK, Ahmed ML, Emmett MA, Preece MA, Dunger DB: Associations between postnatal catch-up growth and obesity in childhood: prospective cohort study. BMJ 2000;320:967–971.
18 Stettler N, Zemel BS, Kumanyika, Stallings VA: Infant weight gain and childhood overweight status in a multicenter, cohort study. Pediatrics 2002;109: 194–199.
19 Jaquet D, Gaboriau A, Czernichow P, Ley-Marchal C: Insulin resistance early in adulthood in subjects with intrauterine growth retardation. J Clin Endocrinol Metab 2000;85:1401–1406.

20 Ong KK, Potau N, Petry CJ, Emmett PM, Sandhu MS, Kiess W, Hales CN: Insulin sensitivity and secretion in normal children related to size at birth, postnatal growth, and plasma insulin-like growth factor-I levels. Diabetologia 2004;47:1064–1070.

21 Dunger DB, Ahmed ML, Ong KK: Early and late weight gain and the timing of puberty. Mol Cell Endocrinol 2006;254:140–145.

22 Persson I, Ahlsson F, Ewald U, Tuvemo T, Qingyuan M, Van Rosen D, Proos L: Influence of perinatal factors on the onset of puberty in boys and girls: implication for interpretation of link with risk of long term diseases. Am J Epidemiol 1999;150:747–755.

23 Luo ZC, Albertsson-Wikland K, Karlberg J: Length and body mass index at birth and target height influences on patterns of postnatal growth in children born small for gestational age. Pediatrics 1998; 102:E72.

24 Leger J, Levy Marchal C, Boch J, Pinet A, Chevenne D, Porquet D, Collin D, Czernichow P: Reduced final height and indications for early development of insulin resistance in a 20 year old population born with intrauterine growth retardation. Br Med J 1997;315:341–347.

25 Boonstra VH, Mulder PGH, De Jong FH, Hokken-Koelega ACS: Serum DHEAS levels and pubarche in short children born small for gestational age (SGA) before and during growth hormone treatment. J Clin Endocrinol Metabol 2004;89:712–717.

26 Jaquet D, Leger J, Chevenne D, Czernichow P, Levy-Marchal C: Intrauterine growth retardation predisposes to insulin resistance but not to hyperandrogenism in young women. J Clin Endocrinol Metab 1999;84:3945–3949.

27 Boonstra VH, Van Pareren Y, Mulder PGH, Hokken-Koelega ACS: Puberty in growth hormone-treated children born small for gestational age (SGA). J Clin Endocrinol Metabol 2003;88: 5753–5758.

28 Lazar L, Pollak U, Kalter-Leibovici O, Pertzelan A, Phillip M: Pubertal course of persistently short children born small for gestational age (SGA) compared with idiopathic short children born appropriate for gestational age (AGA). Eur J Endocrinol 2003;149: 425–432.

29 Hernàndez MI, Martinez A, Capurro T, Pena V, Trejo L, Avila A, Salazar T, Asenjo S, Iniguez G, Mericq V: Comparison of clinical, ultrasonographic, and biochemical differences at the beginning of puberty in healthy girls born either small for gestational age or appropriate for gestational age: preliminary results. J Clin Endocrinol Metabol 2006;91:3377–3381.

30 Ibáñez L, Potau N, Francois I, de Zegher F: Precocious pubarche, hyperinsulinism and ovarian hyperandrogenism in girls: relation to reduced fetal growth. J Clin Endocrinol Metab 1998;83:3558–3562.

31 Ibáñez L, Potau N, Marcos MV, de Zegher F: Exaggerated adrenarche and hyperinsulinism in adolescent girls born small for gestational age. J Clin Endocrinol Metab 1999;84:4739–4741.

32 Ibáñez L, Ong K, De Zegher F, Marcos MV, Del Rio L, Dunger DB: Fat distribution in non-obese girls with and without precocious pubarche: central adiposity related to insulinaemia and androgenaemia from prepuberty to postmenarche. Clin Endocrinol 2003;58:372–379.

33 Ibáñez L, Jaramillo A, Enriquez G, Miro E, Lopez-Bermejo A, Dunger D, De Zegher F: Polycystic ovaries after precocious pubarche: relation to prenatal growth. Hum Reprod 2007;22:395–400.

34 Ibáñez L, Ferrer A, Marcos MV, Hierro FR, De Zegher F: Early puberty: rapid progression and reduced final height in girls with low birth weight. Pediatrics 2000;106:1–3.

35 Ibáñez L, Valls C, Marcos MV, Ong K, dunger DB, De Zegher F: Insulin sensitisation for girls with precocious pubarche and with risk for polycystic ovary syndrome: effects of prepubertal initiation and postpubertal discontinuation of metformin treatment. J Clin Endocrinol Metab 2004;89:4331–4337.

36 Ibáñez L, Potau N, Enriquez G, de Zegher F: Reduced uterine and ovarian size in adolescent girls born small for gestational age. Pediatr Res 2000;47:575–577.

Prof. Anita C.S. Hokken-Koelega, MD, PhD
Division of Endocrinology, Department of Paediatrics
Sophia Children's Hospital, Erasmus University Medical Center
PO Box 2060, NL–3000 CB Rotterdam (The Netherlands)
Tel. +31 10 463 6744, Fax +31 10 463 6811, E-Mail a.hokken@erasmusmc.nl

Kiess W, Chernausek SD, Hokken-Koelega ACS (eds): Small for Gestational Age. Causes and Consequences.
Pediatr Adolesc Med. Basel, Karger, 2009, vol 13, pp 134–147

Neurological and Intellectual Consequences of Being Born Small for Gestational Age

Torsten Tuvemo · Ester Maria Lundgren

Department of Women's and Children's Health, Uppsala University Children's Hospital, Uppsala, Sweden

Abstract

Crude neurological handicaps, like cerebral palsy, are extremely rare in children born small for gestational age (SGA) at term. Such handicaps are more frequent in very premature children. There is seemingly some increase in the risk for nonsevere neurological dysfunction. The effect of SGA birth on intellectual performance is, on the other hand, well documented. Studies in all ages, from the preschool years to adult age, have demonstrated some reduction in intellectual capacity for the group, with a left shift in the distribution of IQ. Those born severely SGA, with small head and/or premature are worse off. Postnatal growth of head circumference and body height influences the outcome, more is generally better. Postnatal undernutrition is detrimental for brain development in this group, but 'high-energy feeding', inducing extreme weight gain, has not demonstrated any advantages compared to 'normal' nutrition. The quality of the food during the first 6–12 months of life seems very important, breastfeeding showing the best outcome, also compared with high-energy formula feeding. There is a possible, but as yet not proven, effect of growth hormone on the intellectual development, especially concerning nonverbal learning capacity. Attention deficit hyperactivity disorder is more frequent in children and adolescents born SGA. In adult age social adaptation is overall normal, but the academic credits and types of profession are slightly less qualified for the group born SGA.

There are today a great number of studies describing the neurological, developmental and cognitive impairment in subjects who were born small for gestational age (SGA). There are both large epidemiological cohort studies and smaller studies on clinical patients. In some studies, SGA children are clearly separated from those with low birth weight due to preterm birth, in others there is a mixture of SGA, with or without verified intrauterine growth restriction, and premature infants born appropriate for gestational age. The outcome has been evaluated at different ages. Relatively few, but some, are longitudinal. The studies will be presented depending on type of problems and age at investigation.

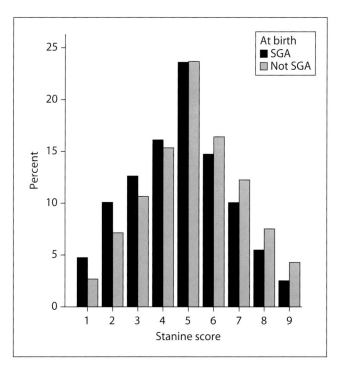

Fig. 1. Distribution of intellectual performance presented as stanine scores in conscripts born SGA (black bars) compared with those born not SGA (grey bars).

Neurological Handicap

Most studies concerning SGA and intellectual functions, exclude infants born with major handicaps such as cerebral palsy and major malformation. However, it has been shown that growth restriction and brain growth are associated with neurodevelopmental deficits [1–5]. Calame et al. [1] compared 50 infants born appropriate for gestational age (AGA) with low birth weight and 33 children born SGA with 41 children born AGA, in studies from Switzerland. Major handicaps such as cerebral palsy and mental retardation were more frequent among those born with low birth weight, but AGA (mean gestational age 29.9 weeks), compared to those born SGA. The authors suggest that gestational age, e.g. preterm birth and perinatal complications, were perhaps more important factors than being born SGA. Neurodevelopmental abilities (visuomotor, fine and gross motor, language and cognitive abilities) were assessed at age 8 years. Neurodevelopmental abnormalities were almost doubled in the 33 children born SGA compared to the 41 AGA children. McCarton et al. [6] described that premature SGA infants (n = 129) were at greater risk for neurodevelopmental impairment than equally premature AGA children (n = 300) at age 6 years in studies from the US. In the Netherlands, 285 children born in the late 1970s were assessed at age 6 years. The children were divided into being born preterm and AGA,

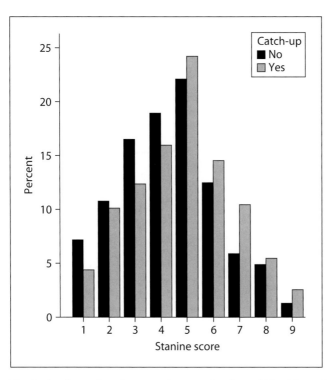

Fig. 2. Intellectual performance in conscripts born SGA with catch-up growth (grey bars) and those without catch-up growth (black bars).

being born SGA at term and being born SGA and preterm. Compared to a control group of children being born AGA at term, both major and minor neurological dysfunctions occurred more frequently in the studied groups. The worst outcome was seen in the group of children born both SGA and preterm [2, 3]. Roth et al. [4] found that at the age of 1 year, one third of SGA and intrauterine growth-retarded (IUGR) term fetuses suffered some, albeit minor, neurological damage. The investigators did not find differences between those with intrauterine growth retardation during the third trimester (IUGR) and those without (SGA). In Canada, a follow-up on preterm infants at the age of 5 years showed that SGA children born before 29 weeks (SGA prematures) had more developmental handicaps than AGA children born that early (AGA prematures). Those with microcephaly had the worst prognosis [5].

The underlying biological mechanisms for the neurodevelopmental deficits in the infants born SGA remain unclear. Cerebral cortex thinning was demonstrated in very-low-birth-weight (VLBW) (mean age 29.1 weeks) and SGA (mean GA 39.5 weeks) children at 15 years of age in studies from Norway [7]. The total brain volume was significantly reduced among SGA births, which according to the authors indicates that intrauterine growth restriction impairs cerebral development. Furthermore, cortical grey matter as a proportion of brain volume was significantly lower in the VLBW, but not in the SGA group compared to AGA controls. More data will probably come up in the future when results from studies on better defined populations will be available.

Intellectual Performance

During the last decades, there has been accumulating evidence to suggest an association between being born SGA and having an increased risk of lowered intelligence, poor academic performance, low social competence, and behavioral problems, compared to individuals born AGA. The evaluation of intellectual performance at early ages is done in several different ways, including standardized tests such as Weschler's Intelligence Scale – Revised (WISC-R) but also teachers' and parents' reports. In adulthood, beside standardized tests, education and occupation are used as indirect variables.

The harmful effects of being born SGA and outcome affecting neurodevelopment and neurological abnormalities must be separated from other risk factors such as being born preterm, perinatal complications and social class. The vast majority of the published studies are done on small cohorts of young children, and not all of them have an adequate control for socioeconomic factors. Moreover, some of the children studied were born in the early years of neonatal intensive care, which might influence the outcome. The results of the studies might also be difficult to interpret, as the definition of SGA varies in the studies and there is a confusion in the literature of being born SGA (statistical description of size at birth at a particular gestational age) and intrauterine growth restriction/retardation, IUGR (failure of growth in utero, due to extrinsic or intrinsic factors).

Preschool Ages

In 2000, Sommerfelt et al. [8] presented a cohort study of 338 term infants born in Norway or Sweden, with birth weights below the 15th percentile. The children were examined at the age of 5 years. Those born SGA had a lower mean IQ compared to those born AGA, even after controlling for confounders such as socioeconomic factors. Similar results have been reported among preschool children born SGA in the early 1980s. IQ scores were lower in 71 SGA children at ages 3, 5, 7 and 9 years compared to 748 normal birth weight children. In addition, the mothers reported more behavioral problems [9].

School Age

Several authors have reported lower IQ, behavioral problems and reduced academic achievement among school age children and young teenagers born SGA compared to controls [9, 10–14]. In studies from Finland, Hollo et al. [11] found that 106 children born with birth weight below the 2.5th percentile had a significantly higher incidence of school failure compared to AGA controls at the age of 10 years. In a similar study from India of 180 children with birth weight less than 2,000 g at term and followed up prospectively for 12 years, it was found that the academic performance was poor compared with controls [15]. This was also found in a study in Canada, in which fetal growth restriction was associated with less favorable academic achievements at age 9–11 years [16]. Even in monozygotic twin pairs with different birth weights, the lighter twins had a lower IQ at 13 years of age. This study indicated that a different fetal environment might have long-term consequences on IQ [17]. In a prospective cohort study, 1,064 SGA children had small but significant deficits in academic achievements at the ages of 5, 10 and 16 years. According to the teachers, SGA children were less likely to be in the top 15% (13 vs. 20%) and had more special education (4.9 vs. 2.3%) [18]. In the United States, the correlation between IQ and birth weight was studied in 3,484 children born at term. A positive correlation was found between IQ and birth weight: each 100-grams of birth weight increase was associated with an increase of 0.5 points of IQ score [19].

Adults

Only a few studies have analyzed intellectual outcome among adults born SGA. In a large study from Israel, including more than 13,000 term infants, it was shown that at the age of 17, the young adults born SGA had a lower IQ compared to those born AGA. However, they did not have a statistically higher risk of low IQ (<85) or lower academic achievement compared to those born AGA after controlling for possible confounding factors [20]. Similar cohort studies from Sweden have shown that low birth weight, short birth length and small head circumference were associated with lower intellectual and psychological performance in 254,426 male conscripts. Boys without catch-up growth in height had the worst outcome [21]. In another Swedish study, 17 SGA subjects were followed prospectively and tested at age 21–28 years. The studied young adults born SGA had lower IQ, particularly verbal IQ, and deficits in figurative learning and memory functions, compared to AGA controls. There were no differences in educational achievement and social adjustment [22]. Strauss et al. [18] showed that at age 26 years, those born with birth weight below the 5th percentile were significantly less likely to have professional or managerial occupations ($p < 0.001$) and were significantly more likely to work as unskilled, semiskilled, or manual laborers than those born AGA ($p < 0.001$), resulting in their reporting a significantly lower income than those with birth weight within normal. These differences persisted after adjusting for sex, social class, and region of birth.

Different Areas of Intellectual Performance

The question has been raised if there are specific cognitive areas in which the individuals born SGA perform lower compared to those born AGA. IQ as a measurement of cognitive functioning is a summary of function in different areas of cognition, and learning deficits have been reported despite normal IQ [23]. Hollo et al. [11] did not find any differences in the intellectual profile among children born SGA at age 10 years. Others have shown lower scores in specific areas. Westwood et al. [24] found lower performance in reading and writing, compared to AGA control subjects, whereas studies by Strauss et al. [18] showed lower grades in mathematics compared to students not born SGA at age 16 years.

In a study on growth-restricted individuals, Ley et al. [25] found a significantly lower verbal and global IQ (96) compared to the control group (102). Interestingly, performance IQ was not significantly affected, at least not to the same extent.

Premature vs. SGA

There are difficulties in separating the effect of being born SGA from being born preterm in the literature. Still, several studies indicate that differences exist [2, 3, 26]. Hutton et al. [26] followed 158 infants born preterm (<32 weeks of gestation) and compared those born AGA with those born SGA. The children were assessed when they were aged 8 or 9 years. The authors conclude that individuals being born preterm and those being SGA differ in their motor and cognitive development. Being born SGA was associated with lower cognitive ability, as measured by IQ scores and reading comprehension. Reduced motor ability was associated with preterm birth. Reading rate and accuracy were not associated with SGA or preterm birth but were socially determined. In a longitudinal Chinese study, Peng et al. [14] found that preterm infants and controls more often had a normal IQ (>85) compared to infants born SGA. The subjects were tested at ages 6 months, 6 years and 16 years.

In summary, being born SGA is found to be associated with moderate school problems and with lower intellectual performance in young adulthood as compared with AGA controls.

Attention Deficit

In a large study from Sweden of 1,480 twin pairs, it was shown that the lighter twin in birth weight-discordant pairs had on average 13% higher ADHD symptom score at age 8–9 years and 12% higher ADHD score at age 13–14 years compared with the heavier twin [27]. SGA adolescents in Norway were found to have more emotional, conduct and attention deficit symptoms [28], but the same authors 'revisiting' attention concluded that the attention problems in teenagers born SGA are limited to a few areas [29].

O'Keeffe et al. [12] followed 7,388 term infants in Australia using the Child Behavior Check List demonstrated that SGA (born with a birth weight <P10) had a modest but independent effect on learning, cognition and attention in adolescence (14 years). Using a cut-off level of the 10th percentile means inclusion of a great part of small normal children, but still the problem of attention deficit seems very limited in the SGA group as a whole. Prediction of risk of attention deficit hyperactivity disorder (ADHD) could be performed using the minor neurological sign test (MNT) according to Sato et al. [30] in a study of LBW infants discharged from neonatal intensive care.

Specific Effects of Growth Restriction

Individuals born SGA are a mixture of children with proved growth restriction, and those who are just small infants born to small parents. The mixture between pathological and just small healthy babies will be different if minus 2 SDS will be used as a cut-off than if the 10th or even the 15th centile is used. In the latter case, the normal small children will probably dominate. Specific diagnosis of growth restriction (previously called growth retardation) is therefore of importance in the analysis of different patient groups. There are few studies starting with findings of reduced placental blood flow or a brake in the fetal growth velocity, and following the children longitudinally. Preeclampsia and twin birth are common reasons for growth restriction, such pregnancies are therefore of interest to study.

Many et al. [31] compared children at age 3 years with intrauterine growth restriction (birth weight <P5), born after pregnancies complicated by preeclampsia (n = 11) with 64 other children, also with a birth weight below the 5th centile. Those with preeclampsia had a lower IQ (mean 85.5) than those without preeclampsia (mean 96.9). In a comparison between 236 twins and 9,832 singletons from Aberdeen born in the 1950s, the twins were found to have a significantly lower IQ than their siblings. The difference was probably at least partly due to growth restriction in the twins [32].

In the study of children born SGA by Paz et al. [20] mentioned earlier, they claim that their SGA children were really growth restricted by controlling for mother's height and BMI, but this is an indirect proof of growth restriction.

More direct signs of an abnormal fetal energy and oxygen support can be obtained by measurement of fetal blood flow. In a large study by Marsal's group [25], 148 children with blood flow velocity waveform studied during pregnancy were evaluated at age 6.5 years. Those with an abnormal wave form had a significantly lower verbal and global IQ (96) compared to the control group (102). In a much smaller study, absent or reverse end-diastolic (ARED) flow velocity in the umbilical artery was a reliable predictor for neurological sequelae, but was not a good predictor of intellectual performance at school at the age of 8.7 years. This was evaluated in 14 IUGR children with ARED versus 11 without [33].

Relation between IQ and Head Circumference

Head circumference is often used as an indirect measure of brain growth in utero. Relations between head circumference (HC), estimated brain weight and development during the first years have been shown [8, 34–36]. In a random cohort, full-scale IQ at 9 years rose with 1.98 points for each SD increase in head circumference at the age of 9 months and 2.87 points for each SD increase at 9 years, after adjustment for many confounders. There was no relation with head size at birth [35]. Studies from Cleveland, Ohio, USA [37] show that small head circumference is an independent risk factor for adverse developmental outcomes as did huge studies from Sweden [21]. Developmental outcome at school age was assessed using age-appropriate cognitive and neuropsychological testing on 128 children born with birth weight below 1,500 g. Subnormal HC was defined as a HC measurement more than 2 SD below the mean for age. At the age of 6 years, 24% of the children born with low birth weight had a subnormal HC, compared to 0% of those born with appropriate birth weight. The children with small head circumference had lower scores on school-age measures of general cognitive ability, language and perceptual motor skills, academic achievement, and adaptive behavior, compared to those with HC within normal. The parents also rated the attention problems higher among these children. Similarly, among 71 children born SGA, Frisk et al. [38] found that small head circumference with poor catch-up head growth after birth was associated with lower academic abilities compared with children born AGA.

Postnatal Growth

The outcome of intellectual performance has been found to be even worse among those who do not catch-up in height, weight and head circumference, compared to those with a catch-up growth [21, 38]. At age 7–9 years poor fetal brain growth in SGA children, followed by little or no postnatal catch-up in brain growth, resulted in the poorest outcomes in IQ scores. SGA children with preserved fetal head growth and good head growth after birth had the best outcomes [38]. In a group of 249 VLBW infants, a head circumference <-2 SDS corrected for GA at the age of 8 months (not at birth) was associated with poor cognitive function, academic achievement and behavior at the age of 8 years [39]. Authors of both papers concluded that brain growth during infancy and early childhood might be more important than during fetal life in determining cognitive function. Among 254,426 young adult males, low birth weight, short birth length and small head circumference were associated with lower intellectual and psychological performance, compared with males born AGA. Lack of catch-up growth in height had the overall worst outcome [21].

Social Adjustment in Adults

There are only few studies on the social outcome of SGA in adult age. This is not unexpected, as in the birth cohorts now in adult ages, newborns were most often grouped together as normal or low birth weight without data on gestational age, birth length, etc. The largest study was performed by Strauss et al. [18], in adults born SGA (n = 1,064) at age 26 years. They could not demonstrate differences in employment, hour per week, marital status or satisfaction with life. SGA subjects had, however, less professional or managerial jobs (8.7 vs. 16.4%) and significantly lower weekly income than AGA individuals. This difference remained after adjustment for confounders.

Neuropsychological outcome, school achievement and social adjustment in young adulthood (21–28 years), was studied in a prospective study in 17 SGA subjects compared to 30 AGA controls. As mentioned above, the adults born SGA had lower IQ, particularly verbal IQ, deficits in figurative learning and memory functions, but there were no differences in educational achievement and social adjustment [22].

As it is known from clinical studies that short SGA children consume significantly less energy, fat and carbohydrates than age-matched AGA controls and significantly less compared to the recommended daily intake, data confirmed by parental reports about a lack of appetite in short SGA children [40], eating disorders have been studied also in adults. There was no increased risk for anorexia nervosa in individuals born SGA, but there was in severely preterm infants in a register study by Cnattingius et al. [41]. In the group of severely preterm the risk seemed higher the lower the birth weight. In a study of the relation between the role of obstetric complications and eating disorder, Favaro et al. [42] found that subjects born SGA had a significantly increased frequency of bulimia nervosa, but not anorexia nervosa. Generally, eating disorder does not appear to be a big problem in this group in adulthood.

Can the Effects of Small-for-Gestational-Age Birth on Intellectual Capacity Be Modified by Feeding Patterns?

So far, the best evidence available for interventions reducing the consequences of being born SGA is for breastfeeding, when it comes to the possibility to modify the negative effects on cognitive development. In a prospective Scandinavian study, Rao et al. [43] demonstrated that 24 weeks of exclusive breast feeding enhanced IQ in SGA children. They followed 220 term SGA children and 299 term AGA children and final outcome was intelligence according to WISC at age 5 years. The group was well off with 98% of the mothers starting breastfeeding, while socioeconomic bias was expected to be small. After controlling for confounding, the results of better IQ, performance but not verbal, was significant in the children which were breastfed exclusively for at least 24 weeks. They could not show any effects of exclusive breast feeding for the first 12 weeks, a fact they discuss from different aspects.

Slykerman et al. [44] studied SGA and AGA children from the Auckland Birthweight Collaborative study. In 531 children of European ethnicity, 223 SGA and 308 AGA, intelligence test scores (Binet) were available. In the SGA group, duration of breastfeeding was positively correlated to intelligence at age 3.5–4 years, which was not the case for the total group. The positive correlation in the SGA group was not limited to exclusive breastfeeding, but included non-exclusive breastfeeding. As this is a large study, the data are important. It was not randomized, but multivariate techniques were used to reduce socioeconomic and other known or suspected confounding. In summary, children born SGA appear to have the greatest advantages of breastfeeding. Are there any hypotheses that could explain the specific advantage of breastfeeding in individuals born growth retarded? In fact there are, as it has been shown that the transfer of certain long polyunsaturated fatty acids is reduced in fetuses with growth restriction [45].

In the study of a high-energy diet by Morley et al. [46], a comparison was also made between breastfeeding and the two formulas, high and normal energy, respectively. The breastfeeding mothers differed highly from those not breastfeeding, but even after controlling for child's gender, birth order, maternal age, education, social class, maternal head circumference and height and whether the mother smoked during pregnancy, there was a highly significant difference in the outcome, Bayleys Scales of Infant Development II at 18 months, in favor of the breastfed children.

In a meta-analysis, Anderson et al. [47] in 1999 of the 14 studies available at that time fulfilling the criteria of analyzing socioeconomic confounding, the result was clear especially in low birth weight infants; an advantage for breast feeding. They did not separate infants born SGA from other patients with low birth weight. In a 'critical review', Jain et al. [48] discussed how good the evidence is that breastfeeding is linked to better intellectual development. Overall, they accepted that most studies had come to that conclusion, but claimed that most studies were not good enough, and even the meta-analysis mentioned above was not conclusive. However, they did not approach the question of a specific effect in SGA children, which makes their critical conclusions less relevant for this discussion, and their criteria for breastfeeding were very strict, which might be less relevant in the light of the findings by Rao et al. [43].

In a comment on the current literature, Carlo Agostini recently summarized the discussion as follows: 'The most relevant take-home message is that, among full-term babies, those who have been subject to intrauterine growth restriction have the greatest advantage of breast feeding' [49]. We think this is in accordance with most data available for the moment.

High-Energy Feeding

In current neonatal practice, infants born preterm as well as SGA are often fed with nutrient-enriched formula or enrichment in combination with breast milk, a so-called 'aggressive nutrition'. The idea is basically to overcome the undernutrition that might happen in preterm very low birth weight infants. De Curtis and Rigo [50] have discussed the problem of under- versus overnutrition, and the different potential harmful effects in SGA children. There is a competition between risks of rapid weight gain and reduced neurodevelopment due to undernutrition. They conclude that human milk after hospital discharge is probably most beneficial for subsequent development.

In a study on children born between 1967 and 1978, Brandt et al. [51] claim that high-energy nutrient intake is necessary for head circumference catch-up growth, but the groups given the different energy intakes were not comparable. Better postnatal head growth might, in turn, result in a better IQ at the age of 7–9 years [38]. However, the relation between increasing head circumference after birth and development of intellectual capacity is far from absolute. In the randomized trial on high-energy feeding comparing two formulas, Morley et al. [46] showed that nutrient-enriched formula improved head growth in SGA infants, but not their neurodevelopment. Head circumference increased more in the high-energy-fed group, especially in girls, and despite this probably increased brain growth, the neurodevelopment scores at 9 months were lower in that group. Their conclusion is that in SGA infants there is no neurodevelopmental advantage from receiving a nutrient-enriched formula.

How can data on head growth, indicating a positive relation to intellectual capacity, fit with the study above? It seems that in studied groups with a mixture of feeding patterns, the importance of head growth is strong enough to be significantly related to IQ. If, on the other hand, those with adequate breastfeeding are separated as a group, the positive effects of breastfeeding are even stronger, and may take over the significance. Thus, quality seems more important than quantity.

Effects of Growth Hormone Treatment on Cognition and Behavior

In a large prospective, longitudinal study the group of Hokken-Koelegas followed 79 short SGA children aged 7.5 years at the start, till final height or near final height [13]. In 53 of them, data on intellectual performance, as measured by short variants of WISC or WAIS, were available, both at the start and during and after long-term growth hormone treatment. The children either were given a 'low dose' or a 'high dose' of growth hormone, i.e. 1 or 2 mg/m^2/day. For ethical reasons, it was not possible to have a control group. Both groups increased their performance IQ and their total IQ significantly during and after treatment, but their not verbal IQ. The two dose groups did not differ in their improvement. The improvement in IQ was related to the degree of catch-up in head circumference but not to the increase in height SDS. This group of children experienced behavioral problems, externalizing, but not internalizing ones. During treatment, the problems were normalized in the two dosage groups in parallel. Self-perception was also significantly improved.

The authors discuss the possible mechanisms for all these changes, but as this was the first, and very controversial, finding in this area, they did not speculate too far, but suspect, of course, that this might be a direct effect of the GH-IGF system on the developing brain. Just recently, the first 2-years results from the Belgian Study Group have been published. They randomized 34 short SGA children into a treatment group and an untreated control group. They could not show any improvement in the treated group [52]. As long as no long-term, randomized, controlled trials are published, we have to be restrictive in the conclusions according GH effects on the SGA brain, but it would be clearly reasonable to expect some positive findings in the future.

References

1 Calame A, et al: Neurodevelopmental outcome and school performance of very-low-birth-weight infants at 8 years of age. Eur J Pediatr 1986;145:461–466.

2 Hadders-Algra M, Touwen BC: Body measurements, neurological and behavioural development in six-year-old children born preterm and/or small-for-gestational-age. Early Hum Dev 1990;22:1–13.

3 Hadders-Algra M, Huisjes HJ, Touwen BC: Preterm or small-for-gestational-age infants: neurological and behavioural development at the age of 6 years. Eur J Pediatr 1988;147:460–467.

4 Roth S, et al: The neurodevelopmental outcome of term infants with different intrauterine growth characteristics. Early Hum Dev 1999;55:39–50.

5 Bardin C, Piuze G, Papageorgiou A: Outcome at 5 years of age of SGA and AGA infants born less than 28 weeks of gestation. Semin Perinatol 2004;28:288–294.

6 McCarton CM, et al: Cognitive and neurologic development of the premature, small for gestational age infant through age 6: comparison by birth weight and gestational age. Pediatrics 1996;98(6 Pt 1):1167–1178.

7 Martinussen M, et al: Cerebral cortex thickness in 15-year-old adolescents with low birth weight measured by an automated MRI-based method. Brain 2005;128(Pt 11):2588–2596.

8 Sommerfelt K, et al: Cognitive development of term small for gestational age children at five years of age. Arch Dis Child 2000;83:25–30.

9 Silva PA, McGee R, Williams S: A longitudinal study of the intelligence and behavior of preterm and small for gestational age children. J Dev Behav Pediatr 1984;5:1–5.

10 Larroque B, et al: School difficulties in 20-year-olds who were born small for gestational age at term in a regional cohort study. Pediatrics 2001;108:111–115.

11 Hollo O, et al: Academic achievement of small-for-gestational-age children at age 10 years. Arch Pediatr Adolesc Med 2002;156:179–187.

12 O'Keeffe MJ, et al: Learning, cognitive, and attentional problems in adolescents born small for gestational age. Pediatrics 2003;112:301–307.

13 van Pareren YK, et al: Intelligence and psychosocial functioning during long-term growth hormone therapy in children born small for gestational age. J Clin Endocrinol Metab 2004;89:5295–5302.

14 Peng Y, et al: Outcome of low birthweight in China: a 16-year longitudinal study. Acta Paediatr 2005;94:843–849.

15 Chaudhari S, et al: Pune low birth weight study: cognitive abilities and educational performance at twelve years. Indian Pediatr 2004;41:121–128.

16 Low JA, et al: Association of intrauterine fetal growth retardation and learning deficits at age 9 to 11 years. Am J Obstet Gynecol 1992;167:1499–1505.

17 Henrichsen L, Skinhoj K, Andersen GE: Delayed growth and reduced intelligence in 9–17 year old intrauterine growth retarded children compared with their monozygous co-twins. Acta Paediatr Scand 1986;75:31–35.

18 Strauss RS: Adult functional outcome of those born small for gestational age: twenty-six-year follow-up of the 1970 British Birth Cohort. JAMA 2000;283:625–632.

19 Matte TD, et al: Influence of variation in birth weight within normal range and within sibships on IQ at age 7 years: cohort study. BMJ 2001;323:310–314.

20   Paz I, et al: Term infants with fetal growth restriction are not at increased risk for low intelligence scores at age 17 years. J Pediatr 2001;138:87–91.

21   Lundgren EM, et al: Intellectual and psychological performance in males born small for gestational age with and without catch-up growth. Pediatr Res 2001; 50:91–96.

22   Viggedal G, et al: Neuropsychological follow-up into young adulthood of term infants born small for gestational age. Med Sci Monit 2004;10:CR8–CR16.

23   Fitzhardinge PM: Early growth and development in low-birthweight infants following treatment in an intensive care nursery. Pediatrics 1975;56:162–172.

24   Westwood M, et al: Growth and development of full-term nonasphyxiated small-for-gestational-age newborns: follow-up through adolescence. Pediatrics 1983;71:376–382.

25   Ley D, et al: Abnormal fetal aortic velocity waveform and intellectual function at 7 years of age. Ultrasound Obstet Gynecol 1996;8:160–165.

26   Hutton JL, et al: Differential effects of preterm birth and small gestational age on cognitive and motor development. Arch Dis Child Fetal Neonatal Ed 1997;76:F75–F81.

27   Hultman CM, et al: Birth weight and attention-deficit/hyperactivity symptoms in childhood and early adolescence: a prospective Swedish twin study. J Am Acad Child Adolesc Psychiatry 2007;46: 370–377.

28   Indredavik MS, et al: Psychiatric symptoms in low birth weight adolescents, assessed by screening questionnaires. Eur Child Adolesc Psychiatry 2005; 14:226–236.

29   Kulseng S, et al: Very-low-birthweight and term small-for-gestational-age adolescents: attention revisited. Acta Paediatr 2006;95:224–230.

30   Sato M, et al: Behavioral outcome including attention deficit hyperactivity disorder/hyperactivity disorder and minor neurological signs in perinatal high-risk newborns at 4–6 years of age with relation to risk factors. Pediatr Int 2004;46:346–352.

31   Many A, et al: Neurodevelopmental and cognitive assessment of children born growth restricted to mothers with and without preeclampsia. Hypertens Pregnancy 2003;22:25–29.

32   Ronalds GA, De Stavola BL, Leon DA: The cognitive cost of being a twin: evidence from comparisons within families in the Aberdeen children of the 1950s cohort study. BMJ 2005;331:1306.

33   Valcamonico A, et al: Absent or reverse end-diastolic flow in the umbilical artery: intellectual development at school age. Eur J Obstet Gynecol Reprod Biol 2004;114:23–28.

34   Gross SJ, et al: Newborn head size and neurological status: predictors of growth and development of low birth weight infants. Am J Dis Child 1978;132: 753–756.

35   Gale CR, et al: Critical periods of brain growth and cognitive function in children. Brain 2004;127(Pt 2):321–329.

36   Cooke RW: Conventional birth weight standards obscure fetal growth restriction in preterm infants. Arch Dis Child Fetal Neonatal Ed 2006.

37   Peterson J, et al: Subnormal head circumference in very low birth weight children: neonatal correlates and school-age consequences. Early Hum Dev 2006;82:325–334.

38   Frisk V, Amsel R, Whyte HE: The importance of head growth patterns in predicting the cognitive abilities and literacy skills of small-for-gestational-age children. Dev Neuropsychol 2002;22:565–593.

39   Hack M, et al: Effect of very low birth weight and subnormal head size on cognitive abilities at school age. N Engl J Med 1991;325:231–237.

40   Boonstra VH, et al: Food intake of children with short stature born small for gestational age before and during a randomized GH trial. Horm Res 2006;65:23–30.

41   Cnattingius S, et al: Very preterm birth, birth trauma, and the risk of anorexia nervosa among girls. Arch Gen Psychiatry 1999;56:634–638.

42   Favaro A, Tenconi E, Santonastaso P: Perinatal factors and the risk of developing anorexia nervosa and bulimia nervosa. Arch Gen Psychiatry 2006;63: 82–88.

43   Rao MR, et al: Effect of breastfeeding on cognitive development of infants born small for gestational age. Acta Paediatr 2002;91:267–274.

44   Slykerman RF, et al: Breastfeeding and intelligence of preschool children. Acta Paediatr 2005;94: 832–837.

45   Cetin I, et al: Intrauterine growth restriction is associated with changes in polyunsaturated fatty acid fetal-maternal relationships. Pediatr Res 2002;52: 750–755.

46   Morley R, et al: Neurodevelopment in children born small for gestational age: a randomized trial of nutrient-enriched versus standard formula and comparison with a reference breastfed group. Pediatrics 2004;113(3 Pt 1):515–521.

47   Anderson JW, Johnstone BM, Remley DT: Breastfeeding and cognitive development: a meta-analysis. Am J Clin Nutr 1999;70:525–535.

48   Jain A, Concato J, Leventhal JM: How good is the evidence linking breastfeeding and intelligence? Pediatrics 2002;109:1044–1053.

49 Agostoni C: Small-for-gestational-age infants need dietary quality more than quantity for their development: the role of human milk. Acta Paediatr 2005;94:827–829.

50 De Curtis M, Rigo J: Extrauterine growth restriction in very-low-birthweight infants. Acta Paediatr 2004; 93:1563–1568.

51 Brandt I, Sticker EJ, Lentze MJ: Catch-up growth of head circumference of very low birth weight, small for gestational age preterm infants and mental development to adulthood. J Pediatr 2003;142:463–468.

52 Lagrou K, et al: Effect of 2 years of high-dose growth hormone therapy on cognitive and psychosocial development in short children born small for gestational age. Eur J Endocrinol 2007;156:195–201.

Prof. emer. Torsten Tuvemo, MD, PhD
Department of Women's and Children's Health
Uppsala University Children's Hospital
SE–751 85 Uppsala (Sweden)
Tel. +46 18 6115926, Fax +46 18 6115583, E-Mail Torsten.Tuvemo@kbh.uu.se

Kiess W, Chernausek SD, Hokken-Koelega ACS (eds): Small for Gestational Age. Causes and Consequences.
Pediatr Adolesc Med. Basel, Karger, 2009, vol 13, pp 148–162

Low Birth Weight in Developing Countries

Maqbool Qadir · Zulfiqar Ahmed Bhutta

Department of Paediatrics and Child Health, The Aga Khan University, Karachi, Pakistan

Abstract

Of the current 4 million estimated deaths in the newborn period, a significant proportion die in association with small size at birth, both prematurity and intrauterine growth retardation (IUGR). In developed
countries, the overwhelming majority of low both weight (LBW) infants are pre-term whereas in developing nations, including those in south Asia, most LBW newborns are full-term infants who are small for gestational age (SGA). Globally, over 20 million LBW infants are born annually, and most of the IUGR-LBW
births are concentrated in 2 regions of the developing world, Asia and Africa, with more than half of LBW
infants in developing countries born in South-Asia alone. As significant proportion of the morbidity and
mortality associated with poor fetal growth is in the neonatal period, and related to birth asphyxia and
infections (sepsis, pneumonia and diarrhea) which account for about 60% of neonatal deaths. Those IUGR-
LBW infants who survive have greatly restricted chances of fully reaching their growth potential. Moreover,
evidence now shows that adults born with LBW face an increased risk of chronic diseases including high
blood pressure, non-insulin dependent diabetes mellitus, coronary heart disease and stroke in adulthood.
There are a number of interventions that work towards reducing the burden of IUGR-LBW. These include
maternal nutrition interventions, increased birth spacing, environmental control with reducing of exposure to smoke and malaria prevention strategies. Although available interventions can make a clear difference in the short term, elimination of stunting will require long-term investments to improve education,
economic status and empowerment of women.   Copyright © 2009 S. Karger AG, Basel

Birth weight is a powerful predictor of infant growth and survival. Infants born with
a low birth weight (LBW) begin life immediately disadvantaged and face extremely
poor survival rates. Of the current 4 million estimated deaths in the newborn period,
a significant proportion die in association with small size at birth, both prematurity
and intrauterine growth retardation (IUGR).

In general, this category includes either preterm infants defined as those infants
born before 37 completed weeks of gestation or those infants born at term but with
features of IUGR, i.e. if sonographic fetal growth falls below the 10th percentile
expected for corresponding gestational age. The term LBW was coined by the World
Health Organization (WHO) to denote newborn infants weighing under 2,500 g (5.5
lb) at birth and as a readily measurable criterion on which to make comparisons

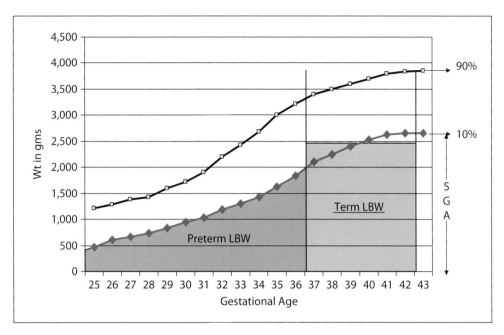

Fig. 1. Classification of newborn infants by birth weight and gestation.

across populations [1]. In developed countries, the overwhelming majority of LBW infants are pre-terms whereas in developing nations, including those in south Asia, most LBW newborns are full-term infants who are small for gestational age (SGA) [2]. SGA are infants whose weight is below the 10th percentile or 2 SD below the mean in growth charts for corresponding gestational age [3, 4]. However, some clinicians define SGA only if the infant birth weight is below the 3rd percentile [5]. Because it is relatively difficult to obtain accurate gestational age assessment across populations, birth weight is used as a reliable metric for comparison. In general, the term LBW is used to denote term growth retarded infants whereas the subcategory very LBW (VLBW) usually reflects the category of preterm infants. Clearly, there can be an overlap between the two categories with some preterm infants also exhibiting features of IUGR whereas some newborn infants with evidence of IUGR may exceed the cut-off of 2,500 g birth weight (fig. 1).

This chapter will henceforth focus on term IUGR-LBW rather than LBW associated with borderline or late prematurity.

Epidemiology and Global Burden

Globally, over 20 million LBW infants are born annually; more than 95% of these are born in developing countries. Most of the IUGR-LBW births are concentrated in 2 regions of the developing world, Asia and Africa, with more than half of LBW infants in

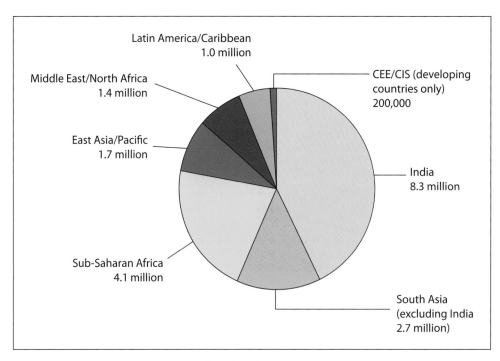

Fig. 2. Distribution of LBW babies, worldwide. Adapted from reference 6 and UNICEF progress for children 2007 [www.unicef.org/progressforchildren/2007].

developing countries are born in South-Asia [6] The highest incidence of IUGR-LBW occurs in the subregion of south central-Asia, where 27% of infants are IUGR-LBW. India alone accounts for 43.6% of IUGR-LBW births in the developing world and more then half of those in Asia (fig. 2). In sub-Saharan Africa, the incidence of IUGR-LBW infants is 13–15%, with little variation across the region as a whole. Central and South America have, on average, much lower rates (10%) while in the Caribbean the level (14%) is almost as high as in sub-Saharan Africa (fig. 3).

The link between IUGR-LBW and stunting in early childhood is unclear. In 2005, 20% of children younger than 5 years in low-income and middle-income countries had a weight-for-age Z score of less than −2.0. The prevalence was highest in south-central Asia and eastern Africa where 33 and 28%, respectively, were underweight. For all developing countries, an estimated 32% (178 million) of children younger than 5 years had a height-for-age Z score of less than −2 in 2005. The largest number of children affected by stunting (74 million) live in south-central Asia [7]. Most of these estimates are at best, calculated estimates from regional studies. To get an actual and accurate estimate from these developing countries is very difficult because of many confounding factors, including lack of resources, and large scattered population residing in rural areas, with a prevalence of home deliveries.

The recent *Lancet* series on maternal and child undernutrition used various procedures to estimate the prevalence of IUGR-LBW [8], which in developing countries

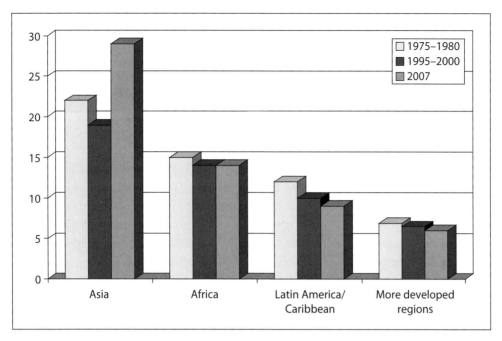

Fig. 3. LBW rates by region over the last three decades. Adapted from reference 6 and UNICEF progress for children 2007 [www.unicef.org/progressforchildren/2007].

accounts for 10.8% of annual live births. The proportions of infants born at term weighing 1,500–1,999 g and those weighing 2,000–2,499 g were estimated using various primary data sets from developing countries, and used to estimate global prevalence of babies born at term, 9.6% weigh 2,000–2,499 g, and 1.3% weigh 1,500–1,999 g globally. The large proportion of infants not weighed at birth constitutes a significant impediment to reliable monitoring of IUGR-LBW also. Many estimates such as the UNICEF data are based on surrogate measures of size at birth. In the developing world as a whole, it is estimated that more than half (58%) of births are not weighed. This proportion is highest in south Asia and sub-Saharan Africa where 74% and 65% of births are not weighed, respectively. Interestingly 58% of babies in the developing world are born with a skilled attendant at delivery, while overall only 42% are weighed [9]. Due to a dearth of female health care providers, only 16% of women seek proper antenatal care and as few as 17% deliver in health facilities [10]. This means that about 83–84% of women go through there pregnancies without any monitoring and deliver babies at home.

It has been a well-recognized fact that a significant proportion of the morbidity and mortality of any neonatal intensive care unit relates to LBW infants, especially those who are preterm. With increasing survival of LBW infants we are learning more and more about there long term outcomes. Poor fetal growth is rarely a direct cause of death, but may contribute indirectly to neonatal deaths, particularly in those due to birth asphyxia and infections (sepsis, pneumonia and diarrhea) which account for about 60% of neonatal deaths. To quantify the risk of neonatal death associated with IUGR-LBW, data from five community-sampled prospective birth cohorts were analyzed (limited to term births and excluding infants <1,500 g birth weight). These analyses indicate that infants born at term weighing 1,500–1,999 g are 8.1 (95% CI 3.3–19.3) times more likely to die, and those weighing 2,000–2,499 g are 2.8 (95% CI 1.8–4.4) times more likely to die from all causes during the neonatal period than those weighing >2,499 g at birth. Based on two studies from South Asia, for deaths due to birth asphyxia, the RR were 5.4 (95% CI 1.8–16.8) for those weighing 1,500–1,999 g and 2.3 (1.3–4.1) for those weighing 2000–2499 g at birth. For infectious causes the RR were 4.2 (95% CI 1.5–11.7) and 2.0 (95% CI 1.2–3.4) for those weighing 1,500–1,999 g and 2,000–2,499 g, respectively.

Infants born with IUGR-LBW suffer from extremely high rates of morbidity and mortality from infectious disease, and are underweight, stunted or wasted beginning in the neonatal period through childhood. Infants weighing 2,000–2,499 g at birth are 4 times more likely to die during their first 28 days of life than infants who weigh 2,500–2,999 g, and 10 times more likely to die than infants weighing 3,000–3,499 g. IUGR-LBW is associated with impaired immune status, poor cognitive development, and high risks of developing acute diarrhea or pneumonia. It is estimated that in Bangladesh, almost half of the infant deaths from pneumonia or diarrhea could be prevented if LBW were eliminated. Those IUGR-LBW infants who survive have greatly restricted chances of fully reaching their growth potential. Moreover, evidence now shows that adults born with LBW face an increased risk of chronic diseases including high blood pressure, non-insulin-dependent diabetes mellitus, coronary heart disease and stroke in adulthood.

Pathogenesis of IUGR-LBW

The important causes of fetal growth restriction and prematurity in developing countries are well known. They include chronic maternal under nutrition (reduced stature), low pre-pregnancy weight or body mass index (BMI), inadequate energy intake and gestational weight gain, cigarette smoking, and specific complications of pregnancy such as genital tract infections, pregnancy-induced hypertension and incompetent cervix [11]. Fetuses with certain genetic or chromosomal disorder are also prone to be intrauterine growth retarded. Table 1 summarizes some of the known risk factors in the pathogenesis of IUGR-LBW [12].

The nutritional status of a woman before and during pregnancy is important for a healthy pregnancy outcome. Aspects of maternal undernutrition, including short stature, and low pre-pregnancy weight or BMI are important risk factors for the delivery of a baby with IUGR. Maternal undernutrition, including chronic energy and micronutrient deficiencies, is prevalent in many regions, especially in South-central Asia where in some countries more than 10% of women 15–49 years old are <145 cm

Table 1. Selected risk factors associated with IUGR-LBW

Demographics
Present low socioeconomic status

Pre-pregnancy
Low weight or height
Short stature
Chronic medical illness
Poor nutrition
Low maternal weight at mother's birth
Previous infant of LBW
Uterine or cervical anomalies
Parity (none or more than five)

Pregnancy
Multiple gestation
Birth order
Anemia
Elevated hemoglobin concentration
Fetal disease
Preeclampsia and hypertension
Infections
Placental problems
Premature rupture of membranes
Heavy physical work
Altitude
Renal disease
Assisted reproductive technology
Exposure to indoor air pollution
Maternal psycho-social stress
Mental health

Behavioral
Low educational status
Smoking
No care or inadequate prenatal care
Poor weight gain during pregnancy
Alcohol abuse
Illicit and prescription drugs
Short inter-pregnancy intervals (less than 6 months)
Age (less than 16 or over 35 years)
Unmarried
Stress (physical and psychological)

tall and maternal wasting, as measured by low body mass index (BMI <18.5) ranges from 10 to 19% in most countries [8]. A serious problem of maternal undernutrition (more than 20% of women with BMI <18.5) is evident in most countries in sub-Saharan Africa, South-central/South-eastern Asia, and in Yemen and countries like

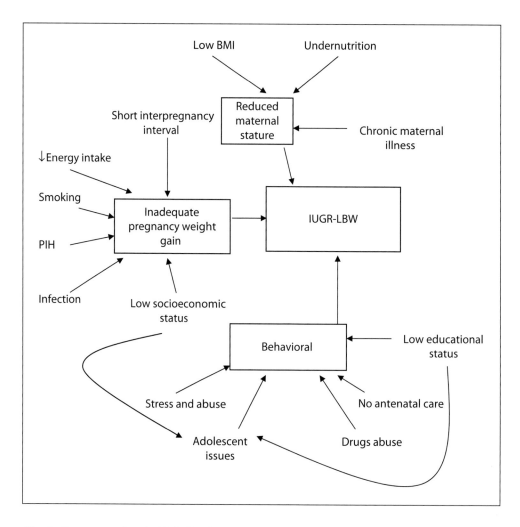

Fig. 4. Factors associated with IUGR-LBW.

India, Bangladesh and Eritrea have a critical problem with a prevalence of low BMI around 40% of all women (fig. 4).

In an evaluation of data of 46 national surveys from 36 developing countries, it was found that women of childbearing age with a low BMI were more likely to have an infant that was smaller or LBW than infants born to mothers with either a normal or high BMI [13]. In the WHO multicenter study, pre-pregnancy weight of expectant mothers, and their weight at 20 and 36 weeks of pregnancy were best indicators for predicting delivery of IUGR-LBW infants [14]. For women of below average pre-pregnancy weight, the biggest indicator (OR = 4.0) for IUGR-LBW was provided by the weight they attained at seven lunar months. This indicator was suggested as the most practical instrument in a primary health care setting to screen IUGR-LBW cases that require intervention.

A very young maternal age exerts indirect effects by influencing the mother's height, weight and nutrition. In a report on 242 adolescent pregnancies (10–18 years old) the LBW rates were 67% [15]. Most of these young mothers belong to low socioeconomic groups, which put further burden on them, not only physically but psychologically. Lack of interest in seeking timely medical help, inadequate intake of nutrients and shorter inter-pregnancy intervals creates a vicious cycles resulting in more chances of IUGR babies in subsequent pregnancies.

Maternal overall nutrition does have a considerable effect on fetal growth. In clinical trials where dietary supplements, balanced in energy and protein, were used, showed consistent improvement in fetal growth (32% reduction in SGA) [16]. Increased physical activity by women, like farming or gathering water, is reported to be associated with infants of low weight, smaller head and mid-arm circumference and lower placental weight even after maternal energy and protein intake were controlled [17]. However, information about effects that micronutrients intakes may have had on the observed association was not given.

Effects of malaria on IUGR-LBW in malaria endemic areas were reviewed in several studies. Maternal malaria chemoprophylaxis was associated with higher maternal hemoglobin levels and higher birth weights [18]. Besides malaria, women in least developed countries are also plagued with many chronic and communicable diseases, which further aggravate their already poor nutritional status. It is known that many maternal factors, such as drug abuse, poor nutritional habits, and cigarette smoking are interrelated and are co variables associated with poor socioeconomic status. Studies of urban air pollution and environmental tobacco smoke (ETS) have shown that several combustion pollutants, including carbon monoxide and small particles, are linked to adverse pregnancy outcomes; still births and LBW. These pollutants are also prominent in smoke from solid fuel used in developing country homes, so there is good reason to expect that this exposure may also impact on pregnancy outcomes in these settings. For % LBW (defined as <2,500 g), one study reported an adjusted odds ratios (OR) for high exposure vs. low exposure of 1.74 (95% CI 1.2, 2.5), one an unadjusted OR of 1.26 (0.77, 2.05), and another OR of 1.20 for term LBW ($p < 0.05$) and 1.50 for pre-term LBW ($p < 0.05$) but results from multivariate analysis were not reported and are presumed non-significant. There was no evidence of statistical heterogeneity among these estimates as all 95% confidence intervals overlap [20]. Meta-analysis of studies from developed countries indicates an association between smoking and exposure to environmental tobacco smoke with IUGR-LBW babies [20]. In an analyses of indoor air pollution and LBW (LBW) in Southern Pakistan, wood fuel was associated with an increased risk of LBW (OR 1.77, 95% CI = 1.2–2.5), after adjusting for other variables [21].

By the same token, maternal psycho-social stress also appears to be an independent risk factor for LBW. Abuse during pregnancy is considered a potentially modifiable risk factor for LBW incidence. The results of a recent meta-analysis indicated that women who reported physical, sexual or emotional abuse during pregnancy were

Table 2. Short-term complications of IUGR-LBW

Effects	
Hyperglycemia	low amount of insulin increased catecholamines
Hypocalcemia	decreased vascular supply in utero
Hypoglycemia	decreased glycogen stores decreased gluconeogeneisis increased sensitivity to insulin
Hypothermia	decreased subcutaneous fat, cold stress large surface area to body weight ratio
Perinatal depression	uterine contractions increase hypoxic stress further decreased cardiac glycogen stores
Depressed immune response	decreased lymphocyte numbers and function decreased immunoglobulin levels increased risk for infections
Polycythemia	chronic hypoxia in utero leading to increased erythropoietin production

more likely than non-abused women to give birth to a baby with IUGR-LBW (OR 1.4, 95% CI 1.1–1.8)[19].

Short Term Complications

Short-term consequences of IUGR-LBW include an increased risk of fetal, neonatal and infant death and impaired postnatal growth, immune function and intellectual development [22] (table 2). The exponential rise in relative risk for neonatal mortality at birth weights below 2.5–3.0 kg is similar in all populations although absolute death rates are considerably higher in developing countries [23].In a study of a cohort of newborns from Bangladesh, neonatal mortality rates for term LBW were reported as 54 deaths per thousand [24] with 84% of all neonatal deaths within the first 7 days, half within 48 h.

Abnormalities have been detected in the polymorphonuclear and lymphocyte cell lines. These abnormalities may persist into childhood, rendering these children more susceptible to infections [25]. Placental insufficiency, leading to poor transfer of antibody, is presumed to be responsible for the abnormally low levels of IgG [26]. Reduction in thymic tissue is noted in 50% of infants leading to reduced number of peripheral T lymphocytes [27]. Attenuated chemotaxis and bactericidal activity have also been described [28]. All these factors could potentially contribute to recurrent infections of both GI and respiratory tracts, right from birth.

In the intermediate term, IUGR-LBW can affect cognitive development by causing direct structural damage to the brain and by impairing infant motor development

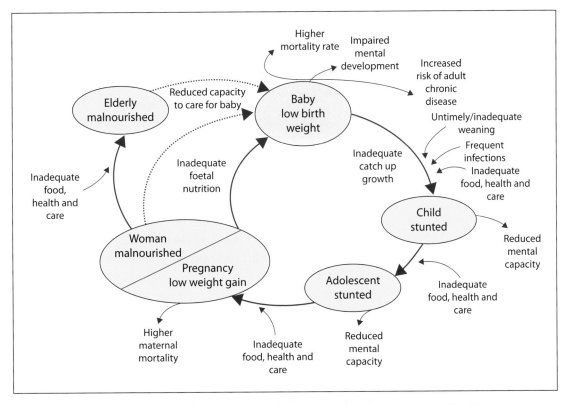

Fig. 5. Nutrition through the life cycle. Source: Commission on the Nutrition challenges of the 21st Century (2000) final report to the ACC/SCN.

[29] and exploratory behavior [30]. Although there are few follow-up studies from childhood to adult age, substantial evidence suggests an association between stunting and present or later cognitive ability or school performance in children, from low-income and middle-income countries (fig. 5) [30].

Long-Term Complications

Long-term consequences also include an increased risk of adult chronic disease (cardiovascular disease and type 2 diabetes) [30, 31] (table 3). This increased risk has been attributed to permanent changes in structure and metabolism resulting from undernutrition during critical periods of early development (the fetal origins of adult disease hypothesis) [32]. An inadequate supply of nutrients forces the fetus to adapt, down-regulate growth and prioritize the development of essential tissues. Adaptations include preferential blood flow to the brain and reduced flow to the abdominal viscera, altered body composition (reduced muscle mass) and reduced secretion of and

Table 3. Long-term complications of IUGR-LBW

Increased risk of adult cardiovascular disease and type 2 diabetes
Increased blood pressure
LBW offspring
Shorter adult height
Reduced mental capacity
Lower attained schooling
Reduced adult income
Developmental delay

sensitivity to the fetal growth hormones (insulin-like growth hormone and insulin). These adaptations enhance immediate survival but may carry a long-term price. An association between LBW and later insulin resistance, a strong risk factor for both cardiovascular disease and type 2 diabetes, is a consistent finding in a number of populations [33]. LBW has also been linked to higher blood pressure in children [34, 35], metabolic syndrome [36] and coronary heart disease in adults [37] in developing countries. The combination of IUGR-LBW followed by obesity in later life appears to carry the greatest risk of insulin resistance. IUGR-LBW also has adverse consequences for future generations. It forms part of an intergenerational vicious cycle of deprivation [38, 39]. For example the poor postnatal growth of IUGR-LBW girls increases their own risk of producing LBW infants.

Birth weight is positively associated with cognitive skills in children, but the effect of environmental factors weakens this association over time. Longitudinal data from developing countries showed that stunting between 12 and 36 months of age predicts poorer cognitive performance and/or lower school grades attained in middle childhood. Longitudinal studies also reveal that IUGR-LBW infants demonstrate retardation in motor, adaptive, personal, social and language development in the first year of life [40]. Some of the developmental retardation can be argued to be the result of continued socio-environmental deprivation. However, even after controlling for socio-demographic traits, it was found in one study that IUGR-LBW affected infants were at greater risk for specific chronic health conditions like mental retardation, developmental delay, Down syndrome, cerebral palsy, autism, muscular dystrophy, cystic fibrosis, sickle cell anemia, diabetes mellitus, arthritis, congenital heart disease, asthma or trouble seeing or hearing [41]. Many studies have shown that mean psychometric or academic test scores of children born small for gestational age are lower than those of children with appropriate birth weight for gestational age. Among studies during the first 6 years of life, the majority demonstrated a disadvantage for those born small. In a large British birth cohort study [42, 43], subjects born SGA had significantly lower score on a number of different cognitive tests at ages 8, 11 and 15 years. At age 26 they had poor reading comprehension and were less likely to have professional or managerial jobs. These differences remained after adjustment for social, demographic and fetal or neonatal factors.

Maternal body size is strongly associated with size of newborn children. Undernourished girls tend to become short adults, and thus are more likely to have small children. In the recent *Lancet* series on Maternal and Child Undernutrition, Victora et al. [44] undertook a meta-analysis of all longitudinal cohorts with reported outcomes and reported that poor growth or stunting in the first 2 years of life leads to irreversible damage, including shorter adult height, lower attained schooling, reduced adult income and decreased offspring birth weight. Early growth failure will lead to reduced adult stature unless there is compensatory growth (so-called catch-up growth) in childhood, which is partly dependent on the extent of maturational delay that lengthens the period of growth. Because maturational delays in low-income and middle-income countries are usually shorter than 2 years, only a small part of the growth failure is compensated for.

There is insufficient information about long-term changes in immune function, blood lipids, or osteoporosis indicators. Birth weight is positively associated with lung function and with the incidence of some cancers, and undernutrition could be associated with mental illness. It was also noted that height-for-age at 2 years was the best predictor of human capital and that undernutrition is associated with lower human capital. It was concluded that damage suffered in early life leads to permanent impairment and might also affect future generations. Perhaps the strongest evidence on a potential long term effect of IUGR and food insecurity and adverse long-term outcomes are from Gambia where a cohort born in a hungry season had reduced adult survival related to infectious illnesses [45, 46], although this was not verified in other settings [47].

Prevention of IUGR-LBW

It is largely assumed that the interventions act on their own and their impacts are estimated with this in mind. In reality, however, some interventions might act synergistically, especially in populations with multiple deficiencies. For instance, the effects of iron supplementation on anemia can be substantially increased by co-administration with vitamin A in deficient populations [48]. In a recent review of interventions that make a difference to nutritional outcomes [49] (table 4), the interventions which had the most profound effects on reducing IUGR-LBW babies included the following.

Balanced energy and protein supplementation during pregnancy: In the reviewed 13 studies and a systematic review, which included 6 studies with information on size at birth, pooled estimate showed that this strategy reduced the risk of LBW babies by 32% (relative risk 0.68, 95% CI 0.56–0.84).

Multiple micronutrient supplementation in pregnancy: Supplementation with three or more micronutrients reduces the risk of term LBW babies by 16%. However, multiple micronutrient supplementation did not differ from iron and folic acid supplementation in terms of rates of LBW babies. A pooled analysis of these data with the results of the Cochrane review showed that multiple micronutrient supplementation in pregnancy can reduce the risk of LBW by 0.84 (0.74–0.95).

Table 4. Maternal nutrition interventions that affect IUGR-LBW

Intervention	Prenatal and antenatal	Neonates (0–1 months)	Infants (1–12 months)	Children (12–59 months)
Balanced energy protein supplementation in pregnancy	32% reduction in term intrauterine growth restriction births (RR 0.68, 95% CI 0.56–0.84). 45% reduction in the risk of stillbirths (RR 0.55, 0.31–0.97)			
Supplementation with iron folate or iron	improved micronutrient status (Hg WMD 12 g/l, 2.93–21.07	improved micronutrient status (hemoglobin WMD 7.4 g/l, 6.1–8.7) potential increased risk of death in malarial areas so only recommended for nonmalarial areas as a treatment strategy		
Multiple micronutrient supplements in pregnancy	multiple micronutrient supplementation (defined as supplementation with 3 or more micronutrients) was associated with 39% reduction in maternal anemia compared with placebo or 2 or less micronutrients (RR 0.61, 95% CI 0.52–0.71)	multiple micronutrients supplementation vs. iron folate results in a significant reduction in the risk of LBW births (RR 0.84, 0.74–0.95)	a recent study in Indonesia of multiple micronutrient supplementation vs. iron-folate tabs in over 31,000 women was also associated with a 22% reduction in infant mortality (RR 0.78, 0.64–0.95)	
Insecticide-treated bed nets	pooled estimates indicated a 23% reduction in risk of delivering a LBW infant (RR 0.77, 0.61–0.98), equivalent to reduction in odds of term LBW of 43%			

Intermittent preventive treatment for malaria in pregnant women (with or without insecticide-treated bed nets) is assumed to reduce the risk of term LBW infants by 37% in unprotected women in their first or second pregnancy.

Strategies to reduce smoking and exposure to environmental smoke during pregnancy have also been hypothesized to be associated with increased birth weight and lower rate of term LBW babies (OR = 0.80, 95% CI 0.65–0.98) [20]. However, definitive proof from intervention studies is limited.

Although available interventions can make a clear difference in the short term, elimination of stunting will require long-term investments to improve education, economic status and empowerment of women. Attention to the continuum of maternal and child undernutrition is essential to the attainment of several of millennium development goals and must be prioritized globally and within countries. Countries with a high prevalence of undernutrition must decide which intervention should be given the highest priority, and ensure their effective implementation at high coverage to achieve the greatest benefit. While the evidence of the benefit of maternal nutrition interventions is promising, what is needed is the political will to combat undernutrition in the developing countries and to address social detriments such as poverty, illiteracy and maternal empowerment.

References

1 De Onis M, Habicht JP: Anthropometric reference data for international use: recommendations from a WHO expert committee. Am J Clin Nutr 1996;64: 650–658.

2 Goplan C: LBWs: significance and implications; in: Nutrition in Children: Developing Country Concerns. New Delhi, National Update on Nutrition, 1994, pp 1–33.

3 Arias F: The diagnosis and management of intrauterine growth retardation. Obstet Gynecol 1977; 49:293.

4 Battaglia FC, Lubchenco LO: A practical classification of newborn infants by weight and gestational age. J Pediatr 1967;71:159.

5 Daikkoku NH, Tyson JW, Graf C, et al: Pattern of intrauterine growth retardation. Obstet Gynecol 1979; 54:211.

6 UNICEF and World Health Organization: Low Birth Weight: Country, Regional and Global Estimates. Geneva, World Health Organization, 2004.

7 WHO: WHO Child Growth Standards: Length/Height-for-Age, Weight-for-Age, Weight-for-Length, Weight-for-Height and Body Mass Index-for-Age: Methods and Development. Geneva, World Health Organization, 2006.

8 Black RE, Allen LH, Bhutta ZA, Caulfield LE, de Onis M, Ezzati M, Mathers C, Rivera J, Maternal and Child Undernutrition Study Group: Maternal and child undernutrition: global and regional exposures and health consequences. Lancet 2008;371:243–260.

9 Department of Reproductive Health and Research, World Health Organization, 'Global monitoring and evaluation' August 2004.[http://www.who.int/reproductive-health/global_monitoring/skilled_attendant.html/].

10 World Health Organization: World Health Report. Make Every Mother and Child Count. Geneva, World Health Organization, 2005.

11 Kramer M, Victoria C: LBW and perinatal mortality; in Samba RD, Bloem MW (eds): Nutrition and Health in Developing Countries. Totowa, Humana Press, 2001.

12 Modified from Committee to Study the Prevention of Low Birth Weight. Washington, Institute of Medicine, National Academy Press, 1985.

13 Nestel P, Rutstein S: Defining nutritional status of women in developing countries. Public Health Nutr 2002;5:17–27.

14 Maternal Anthropometry and Pregnancy Outcomes: A WHO collaborative study. Bull WHO 1995;73 (suppl):S1–S98.

15 Kushwaha KP, Rai AK, et al: Pregnancies in adolescents: foetal, neonatal and maternal outcome. Indian Paediatr 1993;30:501–505.

16 Kramer MS: Balanced protein/energy supplementation in pregnancy (Cochrane review). Cochrane Library 2003:CD000032.

17 Shaw GM: Strenuous work, nutrition and adverse pregnancy outcomes: a brief review. J Nutr 2003;133 (suppl):S1718–S1721.

18 Gulmezoglu M, De Onis M, Villar J: Effectiveness of interventions to prevent or treat impaired fetal growth. Obstet Gynecol Surv 1997;52:139–149.

19 Murphy CC, Schei B, et al: Abuse, a risk factor for LBW? A systematic review and meta-analysis. Can Med Assoc J 2001;164:1567–1572.

20 Windham GC, Eaton A, Hopkins B: Evidence for an association between environmental tobacco smoke exposure and birth weight: a meta-analysis and new data. Pediatr Perinat Epidemiol 1999;13:35–57.

21 Siddiqui AR, Gold EB, Yang X, Lee K, Brown KH, Bhutta ZA: Prenatal exposure to wood fuel smoke and low birth weight. Environ Health Perspect 2008;116:543–549.

22 Barker DJP, Fall CHD: The Immediate and Long-Term Consequences of Low Birth Weight. Technical Consultation on Low Birth Weight. New York, UNICEF, 2000.

23 Ashworth A: Effects of intrauterine growth retardation on mortality and morbidity in infants and young children. Eur J Clin Nutr 1998;52:S34–S42.

24 Yasmin S, Osrin D, Paul E, Costello A: Neonatal mortality of low-birth-weight infants in Bangladesh. Bull WHO 2001;79:608–614.

25 Subhani M: Intrauterine growth restriction; in: Intensive Care of the Newborn and Fetuses. New York, Elsevier Mosby, 2005.

26 Ferguson S: Prolonged impairment of cellular immunity in children with intrauterine growth retardation. J Pediatr 1978;93:52.

27 Chandra RK, Matsumura T: Ontogenic development of the immune system and effects of fetal growth retardation. J Perinat Med 1979;7:279.

28 Prokopowicz J, Zioboro J, et al: Bactericidal capacity of plasma and granulocytes in small-for-date newborns. Acta Paediatr Acad Sci Hung 1975;16:267.

29 Pitcher J, Henderson-Smart D, Robindon J: Prenatal programming of human motor function; in Wintour E, Owens J (eds): Early Life Origins of Health and Disease. New York, Springer Science and Business Media, 2006.

30 Commission on the Nutrition Challenges of the 20th century. ACC SCN Report 2000.

31 Barker DJP: Mothers, Babies and Health in Later Life. London, Churchill-Livingstone, 1998.

32 Curhan GC, Willett WC, Rimm EB, Spiegelman D, Ascherio AL, Stampfer MJ: Birth weight and adult hypertension, diabetes mellitus, and obesity in US men. Circulation 1996;94:3246–3250.

33 Newsome CA, Shiell AW, Fall CH, Phillips DI, Shier R, Law CM: Is birth weight related to later glucose and insulin metabolism? A systematic review. Diabet Med 2003;20:339–348.

34 Bavdekar A, Yajnik CS, Fall CHD, Bapat S, Pandit AN, Deshpande V, Bhave S, Kellingray SD, Joglekar C: The insulin resistance syndrome (IRS) in eight-year-old Indian children: small at birth, big at 8 years or both? Diabetes 2000;48:2422–2429.

35 Law CM, Egger P, Dada O, Delgado H, Kylberg E, Lavin P, Tang G-H, von Hertzen H, Shiell AW, Barker DJP: Body size at birth and blood pressure among children in developing countries. Int J Epidemiol 2000;29:52–59.

36 Bhargava SK, Sachdev HS, Fall CH, Osmond C, Lakshmy R, Barker DJ, Biswas SK, Ramji S, Prabhakaran D, Reddy KS: Relation of serial changes in childhood body-mass index to impaired glucose tolerance in young adulthood. N Engl J Med 2004;350:865–875.

37 Stein CE, Fall CHD, Kumaran K, Osmond C, Cox V, Barker DJP: Fetal growth and coronary heart disease in South India. Lancet 1996;348:1269–1273.

38 Steketee RW: Pregnancy, nutrition and parasitic diseases. J Nutr 2003;133:1661S–1667S.

39 Ramakrishnan U, Martorell R, Schroeder DG, Flores R: Role of intergenerational effects on linear growth. J Nutr 1999;129(2S suppl):544S–549S.

40 Bhargava SK, Datta I, Kumari S: A longitudinal study of language development in small for dates children from birth to five years. Indian Paediatr 1982;19:123–129.

41 Stein REK, Siegel MJ, Bauman LJ: Are children of moderately LBW at increased risk for poor health? a new look at an old question. Pediatrics 2006;118:217–223.

42 Richard M, Hardy R, et al: Birth weight and cognitive function in the British 1946 birth cohort: longitudinal population based study. BMJ 2001;322:199–203.

43 Strauss RS: Adult functional outcome of those born small for gestational age: twenty-six-year follow-up of the 1970 British birth cohort. JAMA 2000;283: 625–632.

44 Victora CG, Adair L, Fall C, et al: Maternal and child undernutrition: consequences for adult health and human capital. Lancet 2008;371:340–357.

45 Moore SE, Cole TJ, Poskitt EM, Sonko BJ, Whitehead RG, McGregor IA, Prentice AM: Season of birth predicts mortality in rural Gambia. Nature 1997;388:434.

46 Moore SE, Cole TJ, Collinson AC, Poskitt EM, McGregor IA, Prentice AM: Prenatal or early postnatal events predict infectious deaths in young adulthood in rural Africa. Int J Epidemiol 1999;28:1088–1095.

47 Moore SE, Fulford AJ, Streatfield PK, Persson LA, Prentice AM: Comparative analysis of patterns of survival by season of birth in rural Bangladeshi and Gambian populations. Int J Epidemiol 2004;33: 137–143.

48 Zimmermann MB, Biebinger R, et al: Vitamin A supplementation in children with poor vitamin A and iron status increases erythropoietin and hemoglobin concentration without changing total body iron. Am J Clin Nutr 2006;84:580–586.

49 Bhutta ZA, Ahmed Tahmeed Black RE, et al, for the Maternal and Child Undernutrition Study Group: Maternal and child undernutrition. What works? Interventions for maternal and child undernutrition and survival. Lancet 2008;371:417–440.

Prof. Zulfiqar Ahmed Bhutta
Department of Paediatrics and Child Health
The Aga Khan University, Stadium Road, PO Box 3500
Karachi 74800 (Pakistan)
Tel. +92 21 4930 051, Fax +92 21 4934 294, E-Mail zulfiqar.bhutta@aku.edu

Author Index

Subject Index